Exploring GeoAI

Exploring GeoAI

Tools and Workflows

Ismael Chivite
Nicholas Giner
Craig Carpenter

Esri Press
Redlands, California

Esri Press, 380 New York Street, Redlands, California 92373-8100

Printed in the United States of America.
30 29 28 27 26   1 2 3 4 5 6 7 8 9 10

Paperback ISBN: 9781589489080
Hardcover ISBN: 9781589489127
E-book ISBN: 9781589489097

Library of Congress Control Number: 2026936189

For purchasing and distribution options (both domestic and international), please visit esripress.esri.com.

Contents

Acknowledgments

We are especially thankful to Vinay Viswambharan, who served as subject matter expert for this book and provided invaluable guidance and direction.

We are grateful to the following for their contributions to this project: Sydney Wallace, Rami Alouta, Colin Kelly, David Yu, Jie Chang, Lindsay Weitz, Priyanka Tuteja, Lisa Tanh, Hong Xu, and Shengan Zhan.

Special thanks to Bob Booth, Delphine Khanna, and Riley Peake for their dedication to developing and shaping excellent GeoAI learning resources. Our appreciation likewise goes to the Training Services team and to Tristan Cassel for ongoing support. Special recognition is also due to Maryam Mafuri, whose quality assurance efforts were essential to success.

Thank you for reading *Exploring GeoAI: Tools and Workflows*. You can look forward to more books on this evolving topic.

Introduction

Overview of the GeoAI capability

Geospatial artificial intelligence (GeoAI) is the fusion of artificial intelligence (AI) with geospatial data, science, and technology. In ArcGIS®, the GeoAI capability includes tools and models that help users automate geospatial data extraction and enhance their spatial analyses. With the growing availability of remotely sensed imagery, sensors, and other geospatial datasets, GeoAI is becoming increasingly important to organizations looking to improve understanding, find deeper insights in their data, and solve complex geospatial problems.

At its base, GeoAI includes the application of deep learning techniques to automate geospatial data extraction from both structured and unstructured data. For example, deep learning can be used in ArcGIS to extract geospatial information from imagery, 3D, and lidar point cloud data using object detection, pixel classification, object tracking, and change detection techniques. Deep learning and large language model (LLM) integration can also be used to extract geospatial data from unstructured text using a variety of natural language processing techniques.

GeoAI also encompasses machine learning and deep learning techniques for solving spatial problems through the analysis of vector, tabular, and time series data. Examples include pattern detection and clustering, predictive analysis, and spatiotemporal forecasting.

GeoAI tools and techniques are embedded throughout a variety of applications and experiences in ArcGIS. You can train your own custom machine learning and deep learning models using geoprocessing tools and wizards in ArcGIS Pro or train them programmatically using the arcgis.learn module of the ArcGIS API for Python. ArcGIS Enterprise users can train their own models using Deep Learning Studio, a web app that provides a multiuser, project-based environment for teams to collaborate on GeoAI projects in a web browser with no software installation required.

In addition to being able to train your own GeoAI models, you can also use pretrained GeoAI models available in ArcGIS Living Atlas of the World for a wide variety of specific tasks over different geographies. These pretrained models help reduce some of the time and manual effort in collecting training data, as well as the amount of compute resources required to train your own models. They can be used out of the box in ArcGIS Pro, Deep Learning Studio, and Map Viewer, and many of them can be fine-tuned with your training data to specific geographies. Currently, more than 100 pretrained models are available for tasks, including extraction of roads, trees, building footprints, cars, water bodies, solar panels, and swimming pools from imagery, as well as land-cover classification and lidar point cloud classification of buildings, power lines, and trees. Other capabilities include object tracking in video, optical character recognition (OCR), time series forecasting, and many more such tasks. The ArcGIS pretrained models also include vision-language models that integrate third-party AI services from OpenAI and Meta, as well as a seamless integration with Hugging Face Hub for a variety of computer vision and text analysis tasks.

In the following chapters, you'll get hands-on experience with using and fine-tuning ArcGIS pretrained models and training your own GeoAI models for imagery and lidar workflows and for performing predictive spatial analysis. With this guidance, you can explore for yourself the power of GeoAI to yield deeper insights from geospatial data.

How to use this book

About this book

This book has been tested for compatibility with ArcGIS Pro 3.6 and the Deep Learning Libraries for ArcGIS Pro 3.6. This book is designed for beginners to intermediate users. No prior experience with GeoAI or deep learning is required. The tutorials in each chapter demonstrate a workflow in a hands-on environment, and most should take about 45 minutes to complete.

Software requirements and licensing

To perform the tutorials in this book, you'll need the following: ArcGIS Pro 3.6, a web browser to access ArcGIS Online, and installation of the Deep Learning Libraries for ArcGIS Pro 3.6. Earlier software versions may not be fully compatible with the tutorial data and may not operate as described.

Hardware requirements for ArcGIS Pro are available at link.esri.com/EsriPress/ProSysReqs.

Information on software access can be found at esri.com.

Downloading the tutorial data

The tutorial data for this book is available for download from ArcGIS Online, available at link.esri.com/ExploreGeoAI/Data. Download the **Exploring GeoAI - Tutorial Data** item. Unzip the file and move it to your C: drive.

Tutorial data that accompanies this book is covered by a license agreement that stipulates the terms of use.

Resources, feedback, and updates

The GeoAI Help documentation provides comprehensive descriptions of software concepts and tools at link.esri.com/ExploreGeoAI/Help.

Feedback, updates, and collaboration are available at Esri Community, the global community of Esri® users. Post any questions or comments about this book to the Esri Press community at link.esri.com/EsriPress/EPCommunity, or reach out to Esri Press at esripress@esri.com.

Visit the book's web page for additional information at link.esri.com/ExploreGeoAI/Book.

CHAPTER 1

Preparing for deep learning using ArcGIS Pro

Objectives

- Learn how to install Python deep learning libraries.
- Understand basic environment configurations.
- Review common GPU settings.

Introduction

To use the deep learning tools in ArcGIS Pro, you must have the correct deep learning libraries installed on your computer.

In this chapter, you will learn how to install these libraries and check that the installation was successful. You will also learn how to verify various settings and troubleshoot common issues.

If you aren't sure whether you have already installed the deep learning libraries or what version you may have installed, you can follow the steps in tutorial 1-2, under the heading "Check whether you have already installed the deep learning libraries," *and review the version* before jumping into tutorial 1-1.

Tutorial 1-1

Install deep learning libraries

In this tutorial, you'll download the deep learning libraries and install them.

Before installing the deep learning package, take a moment to consider the best way to structure your environments, based on your needs. Regardless of your active environment, the deep learning library will be installed in the default environment, arcgis-py3. Although this is commonly done, if you have concerns related to your system architecture or library, you may want to clone the default environment first to preserve that environment. If you follow this chapter and install the deep learning libraries and subsequently clone the default environment arcgis-py3, the deep learning libraries will be included in the new, cloned environment.

You can always use the Microsoft uninstaller to search for the deep learning libraries and uninstall them. You can also create a new environment with only the default libraries and without the deep learning package by following the steps at the end of this chapter, under the heading "Install a new, default environment."

Note: The version of ArcGIS Pro and the version of the deep learning libraries must be the same. Before downloading and installing the deep learning libraries, update your version of ArcGIS Pro, if necessary.

Download and install the deep learning libraries

The deep learning libraries are stored in a GitHub repository. You will download and install them to your computer.

1. Ensure that the ArcGIS Pro application is closed.

2. In your browser, go to the Deep Learning Libraries Installers for ArcGIS repository at link.esri.com/ExploreGeoAI/Installers.

3. Under **Download**, click **Deep Learning Libraries Installer for ArcGIS Pro 3.6**.

   The folder is downloaded.

4. In File Explorer, go to the **Downloads** folder and locate the **ArcGIS_Pro_36_Deep_Learning_Libraries.zip** file.

5. Right-click **ArcGIS_Pro_36_Deep_Learning_Libraries.zip** and extract it.
6. Open the **ArcGIS_Pro_36_Deep_Learning_Libraries** extracted folder.
7. Double-click the **ProDeepLearning.msi** Windows installer package to install the deep learning libraries.

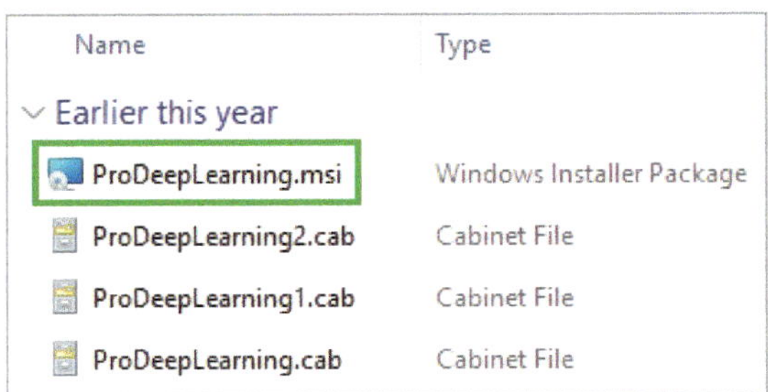

   The installer window appears.

8. Complete the installation.

   The installation may take several minutes.

9. When the installation is complete, click **Finish**.

Review the active environment and verify the installation was successful

Now that you have installed the deep learning libraries, you'll open ArcGIS Pro and verify that the installation was successful.

1. Start ArcGIS Pro and sign in.
2. Click **Settings**.
3. Click **Package Manager**.
4. For **Active Environment**, review the environment name selected.

   In ArcGIS Pro, Python is run by default within a Python environment named arcgispro-py3. You may have created separate environments to install and use different Python libraries. Because the deep learning libraries installer will always install the libraries in the default arcgispro-py3 environment, you will need to ensure the active environment is arcgispro-py3 to use the deep learning tools.

In the following example image, the current active environment is myenv, a custom Python environment created by the user.

You will activate the default environment instead.

5. Open the **Active Environment** list and choose **arcgispro-py3**.

The default Python environment, arcgispro-py3, is now activated.

6. In the **Installed** section, type deep in the search box. Confirm the **deep-learning-essentials** package is listed in the results.

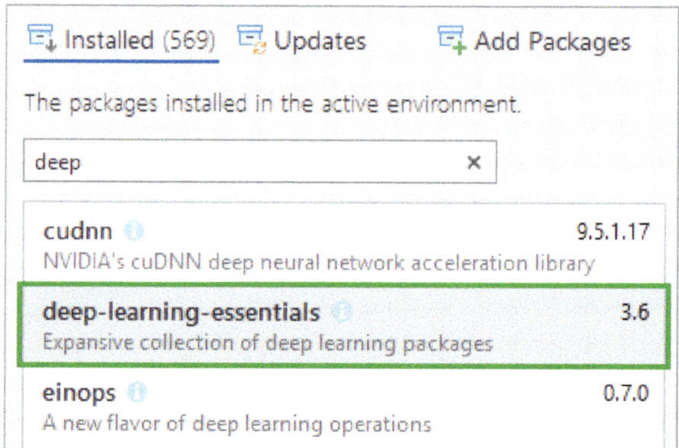

This is one of the required packages for performing deep learning training and inferencing in ArcGIS Pro.

You have now installed the deep learning libraries, activated the correct environment, and confirmed the deep learning libraries have been installed. You are now ready to perform the deep learning workflows presented later in this book.

Tutorial 1-2

Configure additional settings

Installing the deep learning libraries is the most important step to get ready for deep learning in ArcGIS Pro, and it is usually all you need to do. However, in some cases, you may need to review additional settings to ensure that your computer system is configured properly and will be successful running deep learning workflows.

In this tutorial, you will learn about troubleshooting methods for various issues that you may encounter when using the deep learning libraries.

Check whether you have already installed the deep learning libraries and review the version

1. On the Windows taskbar, search for Add or remove programs and press Enter.

   The Installed apps window appears.

2. In the search box, type deep. In the list of results, find **Deep Learning Libraries for ArcGIS Pro**.

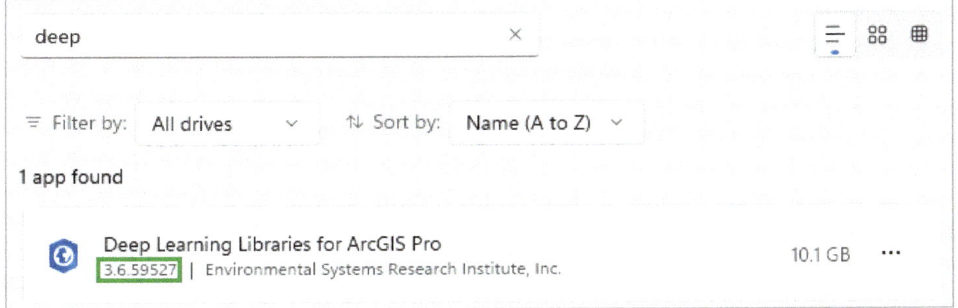

The version number of the libraries appears. The first two digits (in this case, 3.6) indicate the version. These libraries are meant to function with ArcGIS Pro 3.6.

Uninstall an earlier version of the deep learning libraries

If you need a newer version of the deep learning libraries to match your current version of ArcGIS Pro, the safest way to proceed is to uninstall and then install the newer version.

1. In your computer's **Installed apps**, locate the deep learning libraries.
2. Next to the deep learning libraries app, click the **More** options button (three dots) and click **Uninstall**.

   The deep learning libraries are uninstalled.
3. In Windows Explorer, go to **C:\Program Files\ArcGIS\Pro\bin\Python\envs\arcgispro-py3** or the equivalent location for your installation and delete any files still present in it.

   These files may have been left over from a previously modified environment.
4. Follow the tutorial at the beginning of this chapter and install the correct deep learning libraries for the version of ArcGIS Pro that you are using.

   If you still encounter issues with your deep learning workflows, you may need to uninstall the deep learning libraries, manually delete any remaining environments, and uninstall ArcGIS Pro. Then, install the latest version of ArcGIS Pro and the corresponding version of the deep learning libraries.

Check for GPU availability

When running deep learning tools in ArcGIS Pro, it is recommended that you run them using your computer's GPU (graphics processing unit) instead of your computer's standard processor, the CPU (central processing unit). If your computer only has a CPU, you can still use it for simpler deep learning workflows, but the processing time will be longer. Deep learning processing requires a lot of computing power, and GPUs are well adapted because they can perform multiple, simultaneous computations. However, not all GPUs will work for deep learning. More specifically, the deep learning libraries expect an NVIDIA GPU platform. Next, you'll check whether your computer has a suitable GPU.

1. On the Windows taskbar, search for Task Manager and press Enter.
2. In the **Task Manager** window, click the **Performance** tab.
3. In the list of performance indicators, look for a **GPU** option of the **NVIDIA** type.

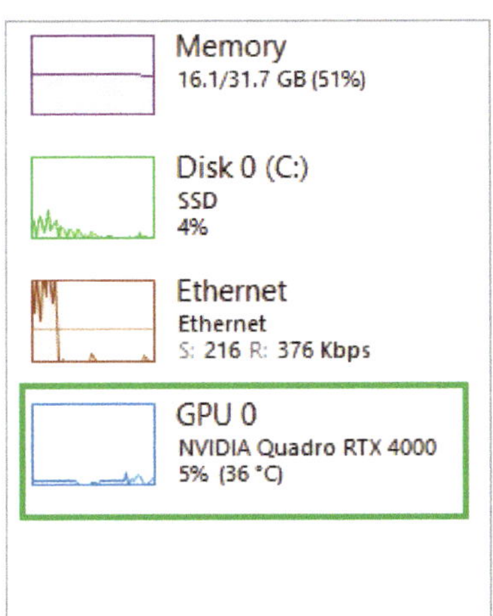

If an NVIDIA GPU is listed, you can run ArcGIS Pro deep learning tools in GPU mode with faster performance. If there is no mention of an NVIDIA GPU, your computer doesn't have that capability. If prompted to choose, you should run ArcGIS deep learning tools in CPU mode.

Your computer might have two GPUs, one NVIDIA and the other one non-NVIDIA. When running a deep learning tool in GPU mode, ArcGIS Pro automatically selects the NVIDIA GPU and ignores the other one.

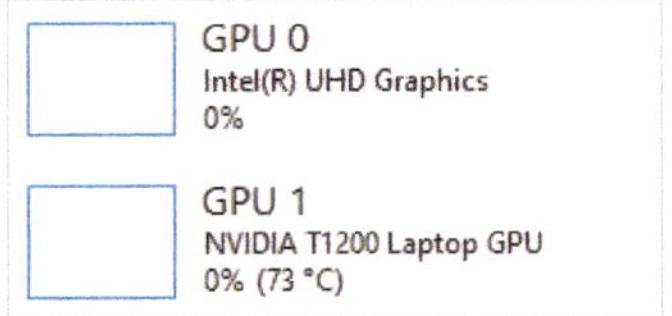

This example shows a computer with two GPUs, including an NVIDIA one.

Note: If you have an older computer, you may have an NVIDIA GPU that won't work with the deep learning libraries.

Check the specifications of your NVIDIA GPU

The main indicator to measure your NVIDIA GPU's power is the amount of dedicated memory. Some 4 to 8 GB of dedicated memory is recommended: 4 GB if running inferencing only and 8 GB for training deep learning models from scratch. You'll check the dedicated memory that your NVIDIA GPU has.

1. In **Task Manager**, click the **NVIDIA GPU** indicator to display more information.

2. Locate the **Dedicated GPU memory** indicator.

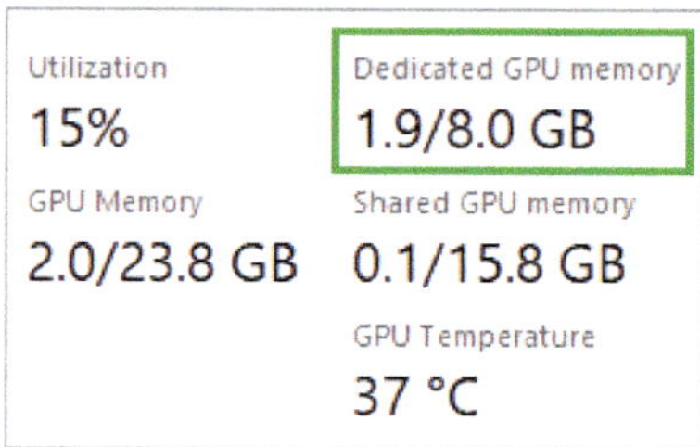

The second number indicates the total amount of dedicated memory available (8 GB in the image shown). The first number indicates how much dedicated memory is currently in use (1.9 GB in the image shown).

3. Locate the **Dedicated GPU memory usage** graph.

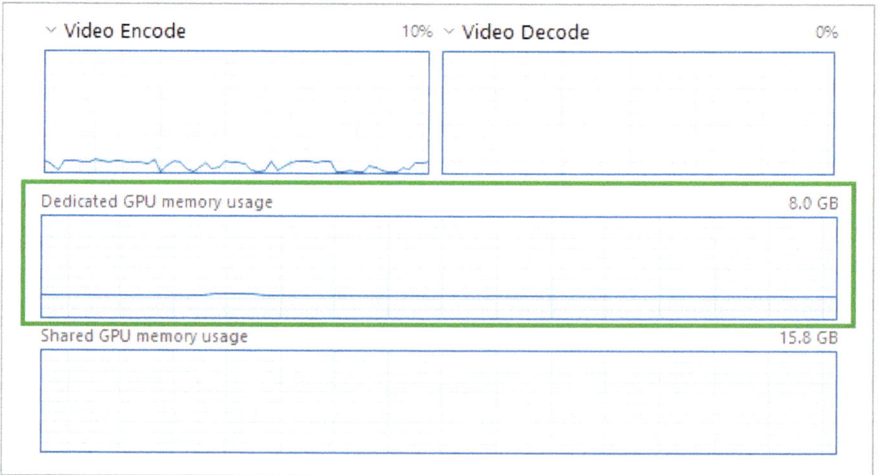

This graph shows the dedicated GPU memory usage in real time. When running a deep learning tool in ArcGIS Pro, you will see the usage level in this graph.

4 Examine the rest of the information provided.

The following specifications are of particular interest:

- The name of the **GPU** model (in the following image, NVIDIA Quadro RTX 4000).
- The **Driver version** (here 31.0.15.3598).
- The **Driver date** (here 5/24/2023).

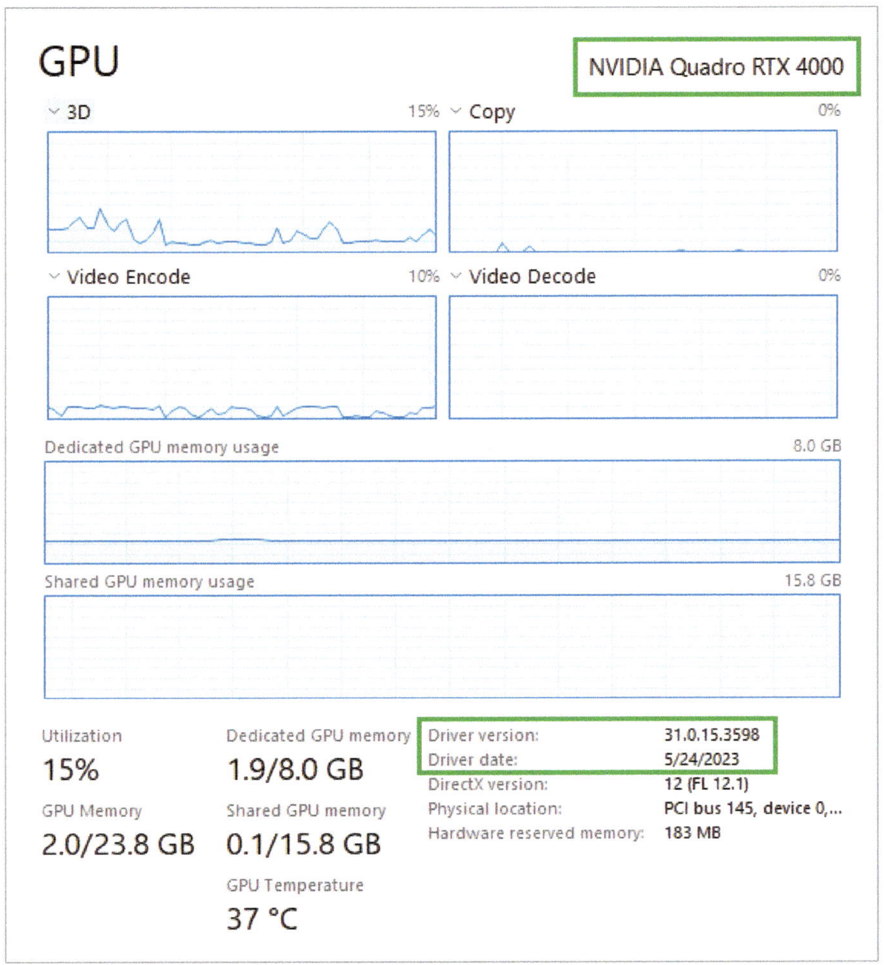

This information will be crucial if you encounter issues due to your GPU driver not being up-to-date.

Update your NVIDIA driver

An out-of-date GPU driver will cause deep learning tools to fail. Check for updates and update your GPU driver if needed.

1. Ensure that ArcGIS Pro is closed.
2. Ensure that you collected the information about your NVIDIA GPU's model and driver version earlier in this tutorial.
3. Go to NVIDIA's Download Drivers page at www.nvidia.com/drivers.
4. In the **Manual Driver Search** section, enter the options that match your GPU.
5. Click **Find**. Find the most recent driver release and click **View**. Then, click **Download**.

   > **Note:** You should usually download the Production Branch/Studio driver.

   The driver's installer file downloads to your computer.
6. On your computer, locate the installer file and install it.

   Your GPU driver is now up-to-date.

Modify the batch size

When running a deep learning tool in ArcGIS Pro, it may run out of dedicated GPU memory and fail. This is generally because the **Batch Size** value you chose is too large for your GPU. To resolve this, you can reduce the **Batch Size** value (for instance, from 4 to 2 or even 1) and run the tool again.

Install a new, default environment

In some cases, you may need an environment that does not include the deep learning libraries—for example, if you want to install and use another Python library but find that some packages conflict. You can create such an environment by using the Python command prompt, as explained in the following instructions.

1. Close ArcGIS Pro.
2. On the Windows taskbar, search for Python Command Prompt and press Enter.

3 In the **Python Command Prompt** window, type the following Conda command:

```
conda create -n myenv arcpy=3.6
```

In this example, `myenv` is the name of the new environment that you are creating, and `3.6` is the version of the ArcPy package, which matches the ArcGIS Pro version installed on your computer (in this case, ArcGIS Pro 3.6).

Note: You can replace **myenv** with a name of your choice.

4 Press Enter.

The command starts running.

5 If asked `Proceed ([y]/n)`, type `y` and press Enter.

Note: The process may take several minutes.

6 When the process is complete, start ArcGIS Pro.

7 Click **Settings**.

8 Click **Package Manager**.

9 For **Active Environment**, review the environment name selected.

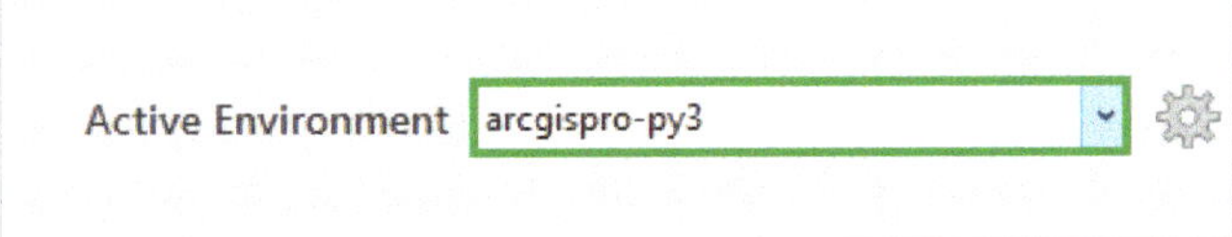

The default environment, arcgispro-py3, is currently active. You'll activate the environment you just created.

10 Open the **Active Environment** list and choose **myenv** (or the other name you chose for your custom environment).

Your new custom Python environment is now active. It doesn't contain the deep learning libraries, but it contains the other (non-deep learning) Python packages you need.

Summary

In this chapter, you learned how to get ready for deep learning in ArcGIS Pro. You considered your preferred environment needs. You installed the Python deep learning libraries and checked that the installation was successful. You then learned how to verify your computer's settings and troubleshoot common issues.

This chapter was based on an Esri tutorial by Bob Booth and Delphine Khanna.

CHAPTER 2

Object detection using pretrained deep learning models

Objectives

- Work with a pretrained model in ArcGIS Pro.
- Refine results to remove false positives and improve accuracy.

Introduction

Deep learning is a powerful approach to detect objects automatically in imagery. A deep learning model is a computer model that is trained using training samples and deep learning neural networks to perform various tasks, such as object detection, pixel classification, change detection, and object classification. With ArcGIS pretrained models, you do not need to invest time and effort into training a deep learning model. The ArcGIS models have been trained on data from a variety of geographies. As new imagery becomes available to you, you can extract features and produce layers of GIS datasets for mapping, visualization, and analysis.

The ArcGIS Living Atlas of the World collection has a variety of pretrained models for anyone to use.

Requirements

- ArcGIS Pro
- ArcGIS Image Analyst extension
- Deep Learning Libraries for ArcGIS Pro
- Recommended: NVIDIA GPU with a minimum of 4 GB of dedicated memory

Tutorial 2-1

Detect boats in aerial images using Text SAM

Text SAM is an open-source sample model that uses free-form text prompts to extract features from imagery. The sample model works by combining Grounding Dino, an open-source vision language model, and the Segment Anything Model (SAM), an image segmentation model.

Note: Read more about the Text SAM model at link.esri.com/ExploreGeoAI/TextSAM.

Text SAM can be used for many use cases. It is a multipurpose GeoAI model that relies on deep learning.

In this tutorial, you will act as an imagery analyst for the City of Copenhagen. You'll detect boats in a marina using the Text SAM model. After running the model in ArcGIS Pro, you'll refine the results.

Set up the project

To get started, you'll download a project that contains all the data for this tutorial and open it in ArcGIS Pro.

Note: To complete this tutorial, you must have the correct deep learning libraries installed on your computer. If you do not have these files installed, save your project, close ArcGIS Pro, and follow the installation steps in chapter 1 first. It is recommended that you have an NVIDIA GPU with a minimum of 4 GB of dedicated memory. If you don't have a GPU, the tool will use your CPU, but it will take longer to process the data. If you are not sure whether your computer has a GPU and what its specifications are, see chapter 1.

1 In File Explorer, open **C:/GeoAI_Data**. Then, open the **Chapter02** folder. Double-click **Boat_Detection.ppkx** to open it in ArcGIS Pro. If prompted, sign in with your ArcGIS account.

A map appears, centered on the Tuborg Havn neighborhood of Copenhagen, Denmark. An image layer, Tuborg_Havn.tif, displays on top of the basemap.

2 Zoom in and pan to examine the imagery.

This aerial imagery was orthorectified to remove any distortions. It is high resolution—each pixel represents a 20-by-20-centimeter square on the ground—and shows boats and other features clearly. It is in the TIFF format with three bands: red, green, and blue, which together form a natural color picture. It has a pixel depth of 8 bits.

Manually identifying all the boats in this image would be time consuming. Instead, you'll use the Text SAM GeoAI model to detect them automatically.

Download the Text SAM model

To use the Text SAM model, you'll first download it to your computer. Text SAM is available on ArcGIS Living Atlas of the World, which is Esri's authoritative collection of GIS data and includes a growing library of deep learning models.

1. In your browser, go to www.livingatlas.arcgis.com.
2. On the ArcGIS Living Atlas home page, in the search box, type Text SAM.
3. In the list of results, click **Text SAM** to open the item page. Review the page.

   Text SAM is a multipurpose model that can be prompted using free-form text prompts to extract features of various kinds from imagery. The output is a polygon layer representing the approximate outline of the objects detected.

   The page also includes useful information on the expected input, which should be 8-bit, 3-band RGB imagery. The model is a good match for the Copenhagen imagery used in this tutorial.

4. Under the **Overview** tab, click **Download**.

   The model file downloads to your computer. The model file is 1.75 GB and may take a few minutes to download.

5. Locate the downloaded **TextSAM.dlpk** file on your computer and move it to a folder where you can easily find it for current and future projects, such as **C:\GeoAI_models**.

> **Note:** You can also use **Text SAM** directly in an ArcGIS Pro geoprocessing tool without saving it first. However, the tool will then download a new copy of the model each time it runs. For that reason, it can save time to store the model locally.

Detect boats using Text SAM

You will now detect the boats present in your Copenhagen image. You'll point the Detect Objects Using Deep Learning geoprocessing tool to the Text SAM model you downloaded, setting that model to be used as one of the parameters.

1. Return to ArcGIS Pro.

2. On the ribbon, on the **View** tab, in the **Windows** group, click **Geoprocessing**.

3. In the **Geoprocessing** pane, in the search box, search for and open the **Detect Objects Using Deep Learning** tool.

4. In the **Detect Objects Using Deep Learning** tool, enter the following settings:

   - For **Input Raster**, select **Tuborg_Havn.tif**.
   - For **Output Detected Objects**, type Detected_Boats.
   - For **Model Definition**, click the browse button. In the **Model Definition** window, browse to the location where you saved the Text SAM model, click **TextSAM.dlpk**, and click **OK**.

Using the ArcGIS Living Atlas portal connection

You can also access the pretrained model without downloading it by using the ArcGIS Living Atlas portal connection in ArcGIS Pro. For the **Model Definition** parameter, click the browse button. In the **Model Definition** window, under **Portal**, click **Living Atlas**. In the search box, type Text SAM (or the pretrained model that you want to use). Select the model and click **OK**.

After a few moments, the model arguments load automatically. If there is a warning symbol next to the model definition, you can ignore it. You will choose a text prompt that corresponds to the objects you want to detect.

5. Under **Arguments**, apply the following settings:

   - For **Text Prompt**, type boat.

     You could add more words to your text prompt separated by commas, such as boat, yacht, canoe. In this case, however, the single-word boat prompt will give excellent results.

- For **Batch Size**, keep the default of **4**.

  When you run the tool, you may get an out of memory error because your computer doesn't have enough memory for the level of processing specified. In that case, try decreasing the Batch Size value from 4 to 2, or even 1.

- For the **NMS Overlap** argument, type 0.7.

  Sometimes the model detects an object more than once. Non Maximum Suppression (NMS) is an optional process that suppresses some of the detected objects when there is duplication. The object that was detected with the highest confidence is kept; the other objects are removed. In the following example image, the boat was detected three times, and with NMS, only one of these three polygons will be kept.

  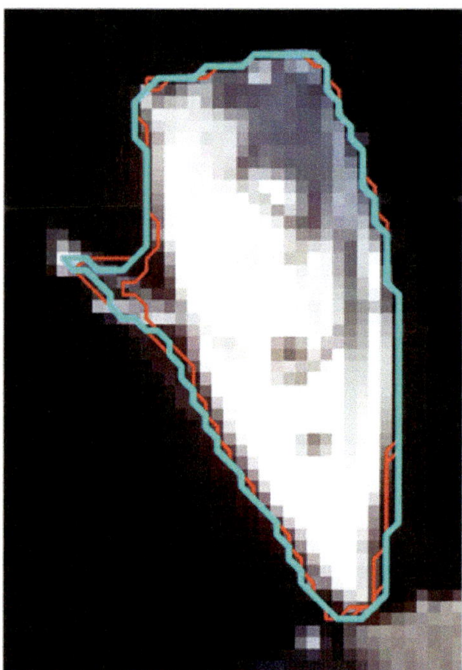

  The NMS Overlap argument determines how much overlap there must be between two detected objects for them to be considered duplicates of each other and for NMS to be applicable. Possible values for that argument are between 0 and 1. For instance, 0.7 means that the overlap should be 70 percent or more.

- Check the box for **Non Maximum Suppression**.

  With the Text SAM model, you can apply NMS during the Text SAM object detection process (that's the NMS Overlap argument) or as a post-processing step (that's the **Non Maximum Suppression** check box). Through trial and error, it was found that the best results for this specific use case are obtained by choosing a high value for the NMS Overlap argument (0.7) and by applying the Non Maximum Suppression post-processing option with its default settings.

  Note: Under **Non Maximum Suppression**, the **Max Overlap Ratio** parameter specifies the overlap for the post-processing NMS step. Just like **NMS Overlap**, it can vary from 0 to 1. The default value of 0 means that as soon as two polygons have an overlap greater than 0, they will be considered duplicates.

- Keep the default values for all the other arguments on this tab.

  If you run the tool as is, it would detect boats in the entire image, which could take 30 minutes to 1 hour based on your computer's specifications. For the brevity of this tutorial, you will detect boats in only a small subset of the input image.

6 On the ribbon, on the **Map** tab, in the **Navigate** group, click **Bookmarks** and select **Detection area**.

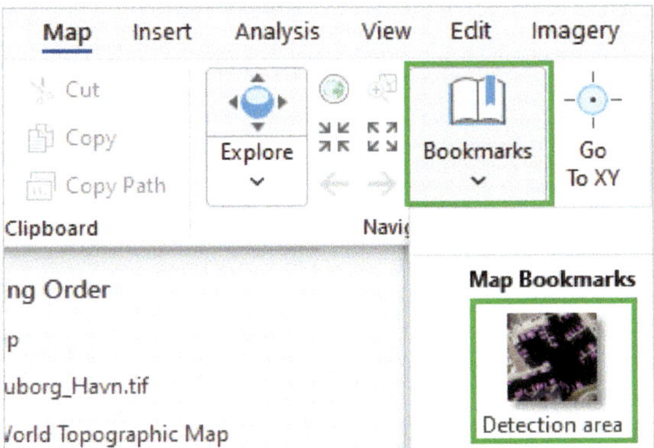

The map zooms in to a smaller area of the Tuborg Havn marina.

7 In the **Geoprocessing** pane, on the **Environments** tab, under **Processing Extent**, click the **Current Display Extent** button.

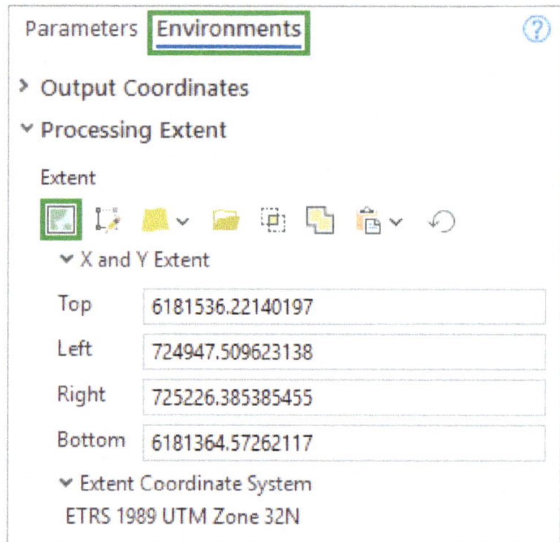

The Top, Left, Right, and Bottom coordinates update to match the current extent showing on the map.

8 Accept all other default values and click **Run**.

You can monitor the progress below the Run button and click View Details to see more information.

After a few minutes, a feature layer, Detected_Boats, appears in the Contents pane and is shown in the map. Each polygon in the layer represents a boat detection by the model.

> **Note:** If you get an out of memory error, try decreasing the **Batch Size** value from 4 to 2, or even 1, and run the process again.

The Text SAM deep learning algorithm is not deterministic, so the results may vary slightly each time you run the tool.

9 In the **Contents** pane, right-click the **Detected_Boats** layer and click **Symbology**.

10 In the **Symbology** pane, click the square next to **Symbol**. Click the **Properties** tab. Style the layer with a red outline and no fill color to better evaluate the results.

You can observe that the model was successful in detecting boats, showing an approximate outline of each boat. However, there are a few cases of false positives—where the model mistakenly identified a boat where there is none, as shown in the following image.

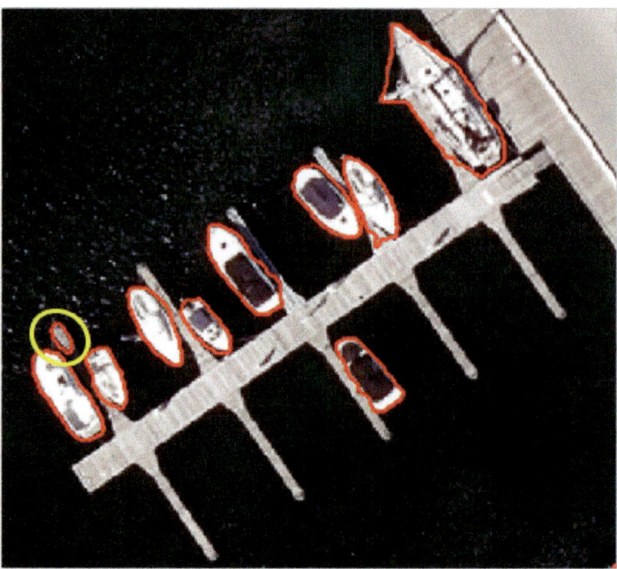

Refine the results

Now you'll refine the results and remove the false positives.

1. In the **Contents** pane, right-click the **Detected_Boats** layer and choose **Attribute Table**.

   In the Detected_Boats attribute table, each row corresponds to a detected boat feature. The number of features you obtained may vary.

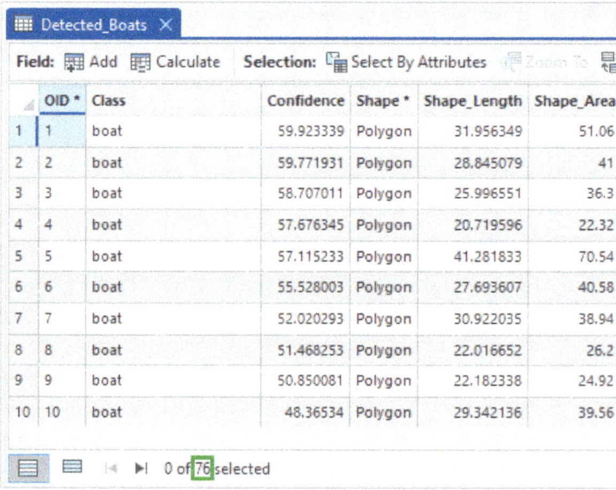

Detected_Boats

Field: Add Calculate Selection: Select By Attributes Zoom To

	OID *	Class	Confidence	Shape *	Shape_Length	Shape_Area
1	1	boat	59.923339	Polygon	31.956349	51.06
2	2	boat	59.771931	Polygon	28.845079	41
3	3	boat	58.707011	Polygon	25.996551	36.3
4	4	boat	57.676345	Polygon	20.719596	22.32
5	5	boat	57.115233	Polygon	41.281833	70.54
6	6	boat	55.528003	Polygon	27.693607	40.58
7	7	boat	52.020293	Polygon	30.922035	38.94
8	8	boat	51.468253	Polygon	22.016652	26.2
9	9	boat	50.850081	Polygon	22.182338	24.92
10	10	boat	48.36534	Polygon	29.342136	39.56

0 of 76 selected

2 Review the following two fields:

- **Confidence**: This field indicates the confidence level with which the model identified each feature as a boat (as a percentage).
- **Shape_Area**: This field indicates the area of each feature (in square meters).

3 Assess the results by sorting these fields. Identify features with small areas or low confidence thresholds that do not represent accurate boat detections.

An area of 9 square meters seems to be the area threshold when boats are accurately identified, and a confidence level of about 28 percent seems to be a good confidence threshold.

4 In the **Contents** pane, right-click **Detected_Boats** and click **Data** > **Export Features**.

5 In the **Export Features** pane, for **Output Feature Class**, type Detected_Boats_Cleaner.

6 Expand **Filter**. Create the following expressions:

- **Where Confidence is greater than 28**
- **And Shape_Area is greater than 9**

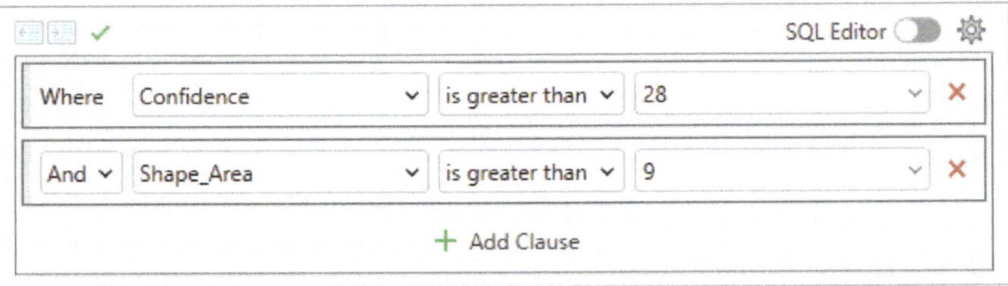

7 Click **OK**.

The new layer is added to the map.

8 In the **Contents** pane, uncheck the box next to **Detected_Boats** to turn the layer off.

9 Change the map's extent to the **Detection area** bookmark.

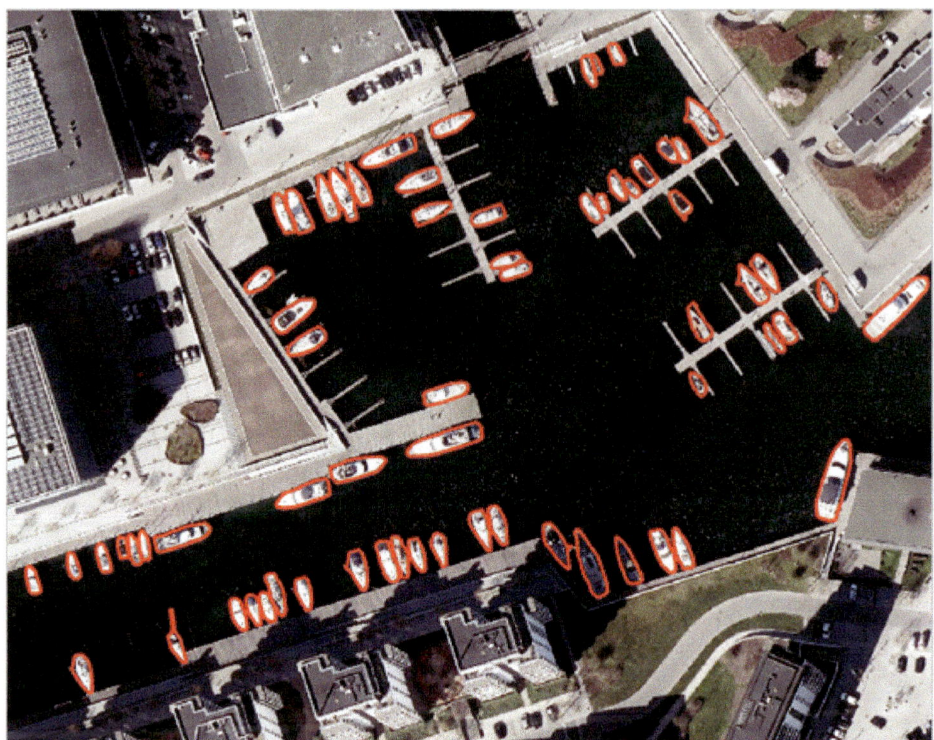

Most of the false positives are now gone.

If you were to use Text SAM to detect boats in every Copenhagen neighborhood, the boat count could be summarized in a graph by neighborhood. The Detected_Boats layer could also be used to create a hot spot map showing the boat concentration levels throughout the city. Finally, this analysis could be repeated regularly on new imagery to identify patterns and change over time.

Best practices for applying Text SAM to imagery

Here are a few tips to successfully apply Text SAM to your data.

Preparing the imagery

The Text SAM model is expecting three-band imagery (red, green, and blue, or RGB). The model also expects the imagery to have an 8-bit pixel depth. If your imagery has a different pixel depth, such as 16 bit, you should convert it to 8 bit.

Finding information about your imagery

If you are not sure what your imagery's properties are (such as number of bands, pixel depth, or cell size), in the **Contents** pane, right-click your imagery layer and choose **Properties**. In the **Properties** pane, click the **Source** pane, and under **Raster Information**, find the **Number of Bands**, **Cell Size X**, **Cell Size Y**, and **Pixel Depth**.

Changing the cell size

If you are not fully satisfied with your first results, you could try varying the **Cell Size** value in the **Detect Objects Using Deep Learning** environment parameters. The cell size (in meters) should be chosen to maximize the visibility of the objects of interest throughout the chosen extent. Consider a larger cell size for detecting larger objects and a smaller cell size for detecting smaller objects. For example, set the cell size for cloud detection to 10 meters, whereas for car detection, set it to 0.3 meters (30 centimeters). While your input image will not change, the tool will resample the data on the fly during processing. For further information regarding cell size, go to link.esri.com/GeoAI/MultiresTextSAM or link.esri.com/GeoAI/PixelSize.

Using a mask

When detecting objects in specific areas of interest, such as boats appearing only in water-covered areas, it can be useful to set a **Mask** in the **Detect Objects Using Deep Learning** environment parameters. A mask is a polygon (or raster) layer that delineates the areas of interest for the analysis—for example, water area boundaries throughout Copenhagen, or perhaps specific marinas if these are the sole targets of your study. When the tool runs, processing will occur only in locations that fall within the mask, saving time and avoiding false positives outside the mask.

Summary

In this tutorial, you downloaded the Text SAM GeoAI model from the ArcGIS Living Atlas website and used it to detect boats in an image. You then used attribute filters to remove false positive features. Finally, you learned a few tips to successfully apply this workflow to your imagery.

This chapter is based on an Esri tutorial by Delphine Khanna.

CHAPTER 3

Improving a deep learning model with transfer learning

Objectives

- Set up a pretrained model with training samples.
- Use transfer learning to fine-tune a model.

Introduction

Deep learning models work best on imagery that is similar to the imagery originally used to train them. If the imagery that you have differs from that original imagery, you can improve the model's performance by giving it examples of the features and imagery in your area of interest. This process is known as transfer learning.

Requirements

- ArcGIS Pro
- ArcGIS Image Analyst
- Deep Learning Libraries for ArcGIS Pro
- Recommended: NVIDIA GPU with a minimum of 8 GB of dedicated memory

Note: Using the deep learning tools in ArcGIS Pro requires that you have the correct deep learning libraries installed on your computer. If you do not have these files installed, first follow the steps in chapter 1 and then continue with this tutorial.

If you are not sure whether your computer has a GPU and what its specifications are, see chapter 1.

Tutorial 3-1

Prepare training samples for transfer learning

In this tutorial, you will generate building footprint layers for various neighborhoods in Seattle, Washington, to support city planning activities. You'll download a pretrained model provided from ArcGIS Living Atlas of the World and prepare your project with training samples to be used for transfer learning.

Set up the project

To get started, you'll download a project that contains all the data for this tutorial and open it in ArcGIS Pro.

1. In File Explorer, open the **C:\GeoAI_Data** folder. Then, open the **Chapter03** folder. Double-click **Seattle_Building_Detection.aprx** to open the project in ArcGIS Pro.

The project opens and imagery of Seattle appears on the map.

2 Zoom in and pan to examine the imagery. Observe that there are many buildings in this image.

Choose a pretrained model

Training a deep learning model from scratch requires feeding it large numbers of examples to show the model what a building is. High-performing models can require tens of thousands of examples. An alternative is to use a model that was already trained for you. You'll retrieve such a model and learn about its specifications.

1 Go to the ArcGIS Living Atlas website at www.livingatlas.arcgis.com.

2 In the search box, type Building Footprint Extraction and press Enter.

The list of results contains pretrained deep learning models for different regions of the world. Since your area of interest is in the United States, you'll choose the model trained on US building footprints.

3. In the list of results, click **Building Footprint Extraction - USA**.

   The description page for the model appears. It contains relevant information about the model. This page notes the type of input the model is expecting. If your input data is not similar enough to the type of data the model was trained on, the model will not perform well.

4. Under **Overview**, click **Download**.

   > **Note:** Ensure that you are signed in to ArcGIS Online with an ArcGIS organizational account.

5. Locate the downloaded file, **usa_building_footprints.dlpk**, on your computer.

6. Move the **usa_building_footprints.dlpk** model file from your download location to the **Pretrained_model** folder, located in the **Chapter03** folder.

Examine the properties of your imagery

Next, you'll investigate how well your data matches the ideal 8-bit, 3-band, high-resolution (10–40 cm) imagery input.

1. Return to your **Seattle_Building_Detection** project in ArcGIS Pro.

2. In the **Contents** pane, right-click **Seattle_RGB.tif** and click **Properties**.

3. In the **Layer Properties** window, click **Source** and expand **Raster Information**.

4. Find the **Number of Bands** field.

   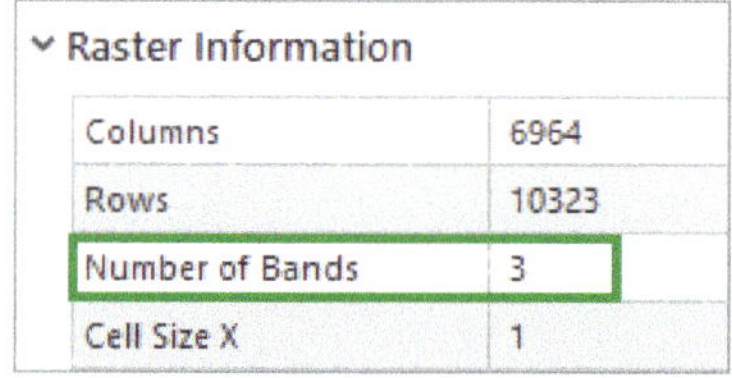

   Your imagery has three bands.

5. Find the **Cell Size X** and **Cell Size Y** fields.

   Cell Size X and Y are both 1.

6 Find the **Pixel Depth** field.

Its value is 8 bit, which matches the 8 bit requested by the model.

Note: The imagery you have been provided came from the NAIP program, which collects multispectral imagery in four spectral bands: red, green, blue, and near-infrared. To obtain the required three bands, the **Extract Bands Raster Function** and **Export Raster** geoprocessing tool were used to save a new layer for use in this workflow. This is an important step you may need to take when working with future projects; if skipped, the model will underperform.

Understand transfer learning

The model expects a higher 10–40 cm resolution, whereas the NAIP imagery was captured at a lower 1 m resolution. If you were to apply the Building Footprint Extraction - USA pretrained model to the Seattle_RGB.tif layer directly, you would get poor results, as shown in the following image.

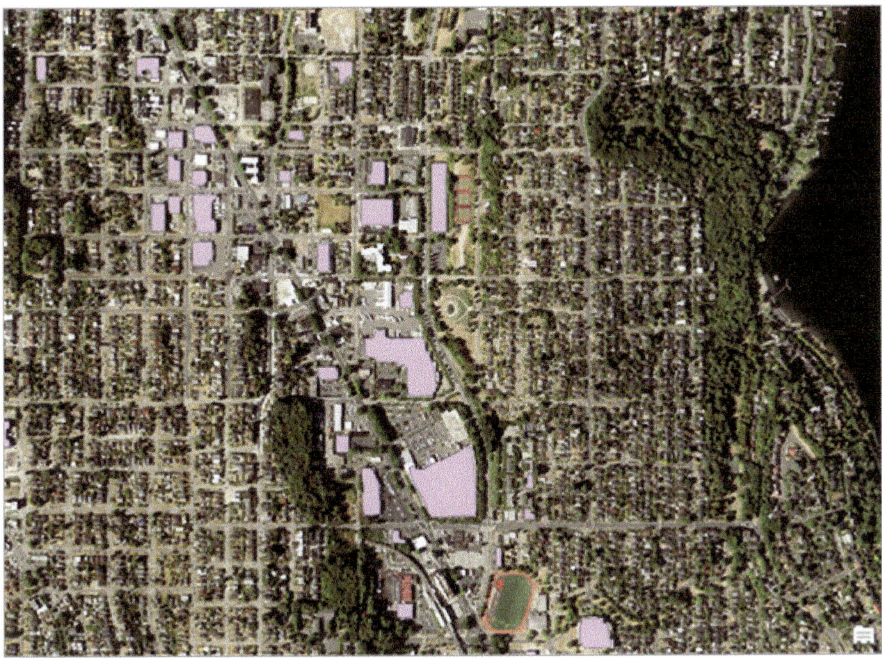

The buildings detected are shown in pink. Because of the resolution mismatch, the model could detect the larger buildings but struggled to identify any of the smaller ones.

One approach to remedy this issue is to use transfer learning. Transfer learning is a technique in machine learning in which knowledge learned from a task is reused to boost performance on a related task. Here, the original task was to detect buildings in 10–40 cm resolution imagery, and the new task is to detect buildings in 1 m resolution imagery. A major advantage of transfer learning is that it requires a relatively small amount of training data and short training time compared with what would be needed to train a model from scratch.

There is a limit to what transfer learning can do if the mismatch between your imagery and the expected input is too extreme. For example, if you had 30 m resolution satellite imagery, where you can barely see the smaller buildings, it is unrealistic to think that the model could be fine-tuned to be successful on that imagery. The more dissimilar the new task is from the original one, the less effective transfer learning will be.

Note: Transfer learning doesn't work on all deep learning pretrained models. For instance, models relying on SAM and DeepForest don't support transfer learning. You can review the description of the pretrained model on the ArcGIS Living Atlas website to see whether it relies on SAM or DeepForest.

Prepare training samples for transfer learning

To perform transfer learning, you first need to produce training examples to show the model what a building footprint looks like in your data. If you were training a model from scratch, you would need tens of thousands of samples. With transfer learning, however, you need only a few hundred. In this part of the tutorial, you'll review the provided training samples and export the feature class and the imagery into training chips used for transfer learning.

In this workflow, Seattle area training samples are provided. If you are interested in applying this workflow to a future project, you will need to create your own feature class of building footprints to use as training samples.

1. In the **Catalog** pane, expand **Databases** > **Output_provided.gdb**.

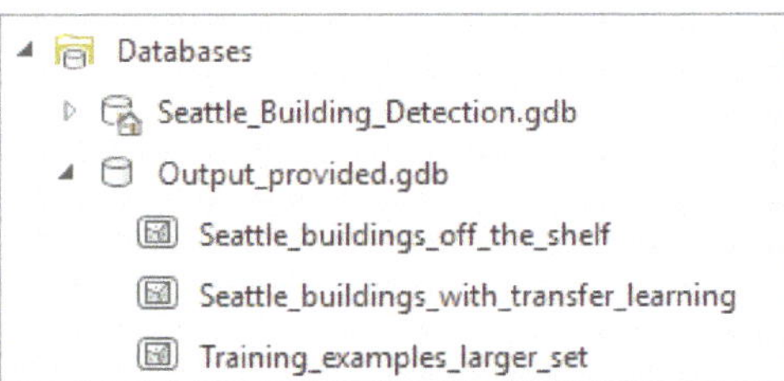

2 Right-click **Training_examples_larger_set** and click **Add To Current Map**.

The building footprints, made up of more than 200 training samples, are added to the map and Contents pane.

3 In the **Contents** pane, right-click the building footprints layer and click **Zoom To Layer**.

In the building footprint feature class, a Class field has been added to designate the features as belonging to a specific class. In some workflows, labeled objects might belong to different classes (or categories), such as building footprints, trees, or cars. In this tutorial, there is only one class: building footprints. It has been populated with a numeric value, arbitrarily represented by the numeric value of 1. Thanks to the Class field, the model will know that all the training examples are the same kind of object: building footprints represented by 1.

Clip the imagery

A deep learning model can't train over a large area in one pass. It can only handle smaller cutouts of the image, known as chips. A chip is made of an image tile and a corresponding label tile, which shows where the objects (in this case, buildings) are located. These chips are fed to the model during the transfer learning training process.

A training chip is shown, with its image tile (*left*) and its corresponding label tile (*right*).

You'll use the Seattle_RGB.tif imagery and the Training_examples layer to generate training chips. It is important to avoid generating chips that contain unlabeled buildings. Having such chips would be the equivalent of showing buildings to the model, while stating that they are not buildings at all. This would be confusing for the model and hurt its performance. To prevent this, you'll create a clip of the imagery that is limited to the extent where the training samples are located.

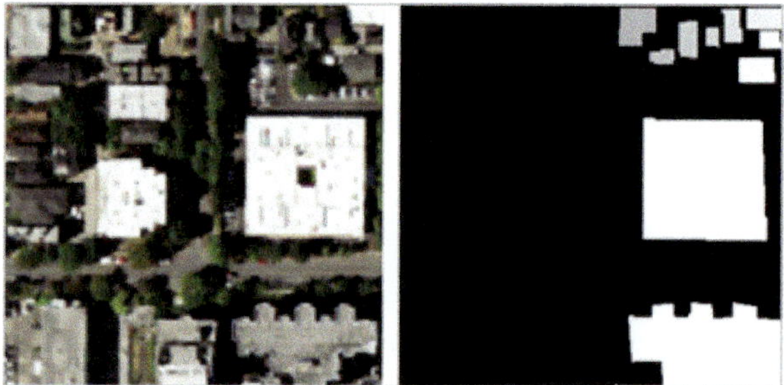

A training chip where some buildings have not been labeled. Such chips must be avoided.

1. On the ribbon, in the **Analysis** group, click **Tools** to open the **Geoprocessing** pane.

2 Search for and open the **Clip Raster** tool and apply the following settings:

- For **Input Raster**, select **Seattle_RGB.tif**.
- For **Output Extent**, select **Training_examples_larger_set**.
- For **Output Raster Dataset**, click the browse button. In the **Output Raster Dataset** window, browse to **Folders** > **Seattle_Building_Detection** > **Imagery_data**. For **Name**, type Seattle_RGB_clip.tif and click **OK**.

3 Click **Run**.

In the Contents pane, the Seattle_RGB_clip.tif layer appears.

4 In the **Contents** pane, uncheck the box next to **Seattle_RGB.tif** to turn the layer off.

On the map, you now see only the clipped layer and the training samples. All the buildings that appear in the imagery have a corresponding building polygon.

Generate training chips

Next, you'll generate the training chips and store the data elements related to the transfer learning process.

1. Search for and open the **Export Training Data For Deep Learning (Image Analyst Tools)** tool and apply the following settings:

    - For **Input Raster**, select **Seattle_RGB_clip.tif**.
    - For **Output Folder**, click the browse button. In the **Output Folder** window, browse to **Folders** > **Seattle_Building_Detection** > **Transfer_learning_data**. For **Name**, type Training_chips, and click **OK**.
    - For **Input Feature Class**, select **Training_examples_larger_set**.

      The chips generated from the clipped imagery and training examples will be stored in a folder named Training_chips.

    - For **Class Value Field**, select **Class**.

      The Class field specifies which objects belong to what labels (in this case, all objects belong to class 1, representing building footprints).

    - For **Tile Size X** and **Tile Size Y**, verify that the value is **256**.

      These parameters decide the size of the chip in the X and Y directions (in pixels). In this case, the default value of 256 is a good choice.

      You want to make your training chips as similar as possible to the chips that were used to train the original model. The original model was trained on 512 × 512 chips produced from 10–40 cm resolution data. Your NAIP imagery is 1 m resolution. A 256 × 256 pixels chip at that resolution will cover roughly the same area as a 512 × 512 chip at 40 cm resolution. So, 256 × 256 is a good chip size to choose.

Note: One way to know the chip size that was originally used in the pretrained model is to look inside the .dlpk file. In File Explorer, make a copy of the **usa_building_footprints.dlpk** file to a separate folder and change its extension from .dlpk to .zip. Right-click the .zip file and extract it. Among the extracted files, locate **usa_building_footprints.emd** and change its extension to .txt. Open **usa_building_footprints.txt** in a text editor and look for the lines **"ImageHeight"** and **"ImageWidth"**.

```
],
"ArcGISLearnVersion": "1.9.0",
"ModelFile": "usa_building_footprints.pth",
"ImageHeight": 512,
"ImageWidth": 512,
"ImageSpaceUsed": "MAP_SPACE",
"LearningRate": "1.0000e-04",
"ModelName": "MaskRCNN",
"backend": "pytorch",
"average_precision_score": {
    "1": 0.7187471585753209
},
```

- For **Stride X** and **Stride Y**, type 64.

  This parameter controls the distance to move in the X and Y direction (in pixels) when creating the next image chips. This value is decided by how much training data you have. You can maximize the number of chips generated by setting this value smaller. You can experiment with this value; however, for this tutorial, a value of 64 was found to work well.

- For **Metadata Format**, select **RCNN Masks**.

  Different deep learning model types require different metadata formats for the chips. Earlier in the workflow, you noted that the pretrained model was based on the MaskRCNN architecture. Here, you must choose the value corresponding to that model.

2 Accept all other default values and click **Run**.

After a few moments, the process is completed.

Examine the training chips

You'll examine some of the chips you generated.

1. In the **Catalog** pane, expand **Folders** > **Seattle_Building_Detection** > **Transfer_learning_data** > **Training_chips**.

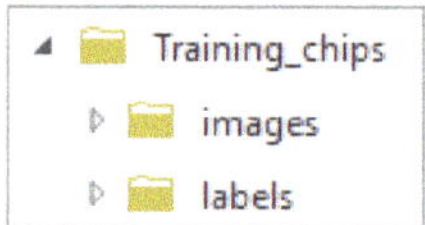

The image tiles are in the images folder and the label tiles in the labels folder.

2. Expand the **images** folder. Right-click the first image, **000000000000.tif**, and click **Add To Current Map**.

Note: If you are prompted to calculate statistics, click **No.**

3. In the **Contents** pane, turn off **Training_examples_larger_set** and **Seattle_RGB_clip.tif** to better see the tile.

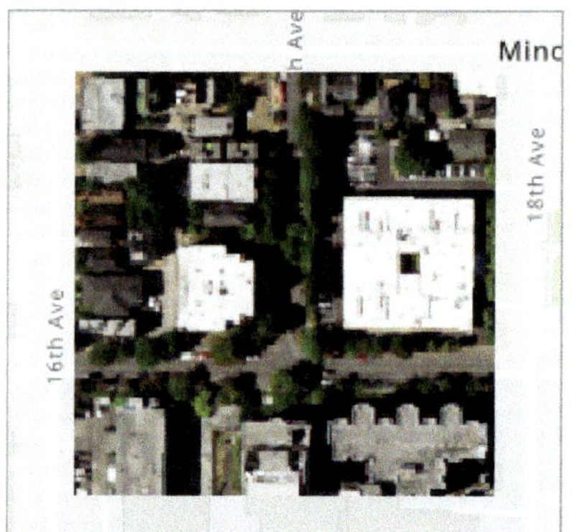

4. In the **Catalog** pane, collapse the **images** folder and expand the **labels** and **1** folders. Add the first label tile, **000000000000.tif**, to the map.

Note: If you are prompted to calculate statistics, click **No.**

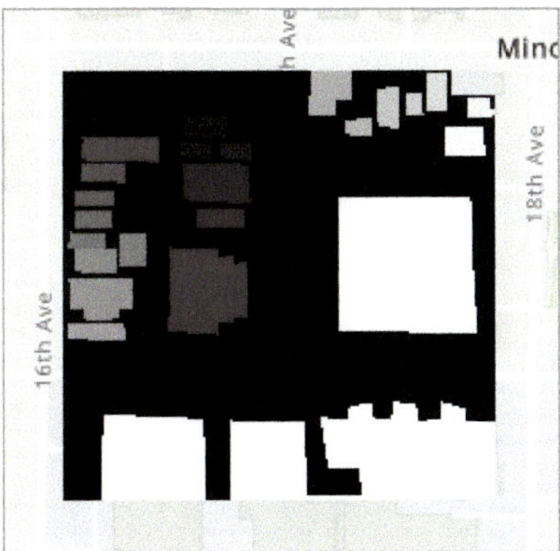

Image and label pairs can be recognized by their identical names.

5 In the **Contents** pane, turn the label tile on and off to reveal the image tile underneath. Then, turn off the image tile.

6 Click some of the label tile pixels to view their values in the pop-up.

Note: On the label tile, the pixels that don't represent a building have the value 0. All pixels that represent a building have a value greater than 0. The specific values come from the object IDs of the original building polygons.

 When finished, remove all the tiles from the **Contents** pane. Turn the **Training_examples_larger_set** and **Seattle_RGB.tif** layers back on.

You generated training chips and you are now ready to start the transfer learning process.

Tutorial 3-2

Conduct transfer learning and extract building footprints

In this tutorial, you'll conduct transfer learning. You'll use the chips you generated to further train the usa_building_footprints.dlpk pretrained model. Then, you'll apply the fine-tuned model to your Seattle imagery.

Fine-tune the model

First, you'll use the Train Deep Learning Model tool to fine-tune the model.

1 In the **Geoprocessing** pane, search for and open the **Train Deep Learning Model** tool and apply the following settings:

- For **Input Training Data**, click the **browse** button. Browse to **Folders** > **Seattle_Building_Detection** > **Transfer_learning_data**. Select **Training_chips** and click **OK**.
- For **Output Folder**, click the browse button. Browse to **Folders** > **Seattle_Building_Detection** > **Imagery_data** > **Transfer_learning_data**. Type Seattle_1m_Building_Footprints_model and click **OK**.
- For **Pre-trained Model**, click the **browse** button. Browse to the folder where you saved the **usa_building_footprints.dlpk** pretrained model, select it, and click **OK**.

Seattle_1m_Building_Footprints_model will be the name of the new fine-tuned model resulting from the transfer learning process.

Note: It is easier to remember what model was trained on which data if you keep each model and its corresponding training chips in the same folder.

2 Expand the **Advanced** section and confirm that the **Freeze Model** box is checked.

The Freeze Model option ensures that only the final layer of the model will be impacted by the new training data, while its core layers remain unchanged. This setting is chosen in many transfer learning cases, as it avoids the risk of the model unlearning its core knowledge.

Note: If you now see an error indicator next to **Input Training Data**, you do not have the correct version of the Deep Learning Libraries installed. Refer to chapter 1. When the installation is complete, you can reopen your ArcGIS Pro project and continue with the tutorial.

3 Expand the **Data Preparation** section and apply the following settings:

- For **Data Augmentation**, select **None**.
- For **Batch Size**, type 4.

Note: If you get an out of memory error, it may be because your computer doesn't have enough memory to process four tiles at a time. Try decreasing the **Batch Size** value from 4 to 2 or 1.

4 Accept all other default values and click **Run**.

The process might take 10 minutes or more to run.

You now have an enhanced model, Seattle_1m_Building_Footprints, that is fine-tuned to better perform on your data.

Run inference

Now that you have completed transfer learning, you'll use your fine-tuned model to run inference on the Seattle_RGB.tif imagery layer and detect the buildings it contains.

1 Search for and open the **Detect Objects Using Deep Learning** geoprocessing tool and apply the following settings:

- For **Input Raster**, select **Seattle_RGB.tif**.
- For **Output Detected Objects**, type Seattle_buildings.

- For **Model Definition**, click the browse button. Browse to **Seattle_Building_Detection** > **Transfer_learning_data** > **Seattle_1m_Building_Footprints_model** and select **Seattle_1m_Building_Footprints_model.dlpk**. Click **OK**.

  As the model definition loads, the model's arguments fill in automatically.

- For **Padding**, confirm that the value is **64**.

  Padding designates a border area in every image chip that will be ignored during detection. If a fragment of a building appears at the edge of an image chip, the padding will ensure that it is not considered for detection. A value of 64 indicates that the padding will be 64 pixels wide on every side of the image chip.

  Note: The model will adjust the stride to match the value of the padding. As the model strides to neighboring areas, a building that appeared as a fragment on the edge of a previous image chip will soon appear in its entirety at the center of one of the next image chips, where it will be successfully detected. Learn more about padding (and other inferencing parameters) at link.esri.com/ExploreGeoAI/DeepLearningTips.

- For **Batch Size**, use the same value as you did for the training process (**4** or less).

- For **Confidence Threshold**, confirm that the value is **0.9**.

  This value is a cutoff value between 0 and 1. It expresses how confident the model must be before it declares an object to be a building. The 0.9 value indicates that the model should have a 90 percent confidence level.

- For **Tile Size**, confirm that the value is **256**.

  This value indicates the size of the imagery chips that the model will take in to run inference. This value should be the same as the size of the chips that were used to train the model.

- For **Non Maximum Suppression**, check the box.

  When there are overlapping building footprint duplicates, the Non Maximum Suppression option ensures that only the building polygon feature with the highest confidence is kept and the other ones are deleted.

At this point, you could run the tool as is; It would proceed to detect buildings over the entire Seattle_RGB.tif image, which could take 10 minutes to one hour based on your computer's specifications. For the brevity of this tutorial, you'll detect buildings in only a small subset of the image input.

2 On the ribbon, on the **Map** tab, in the **Navigate** group, click **Bookmarks** and click the **Inference extent** bookmark.

The map zooms in to a smaller area of Seattle.

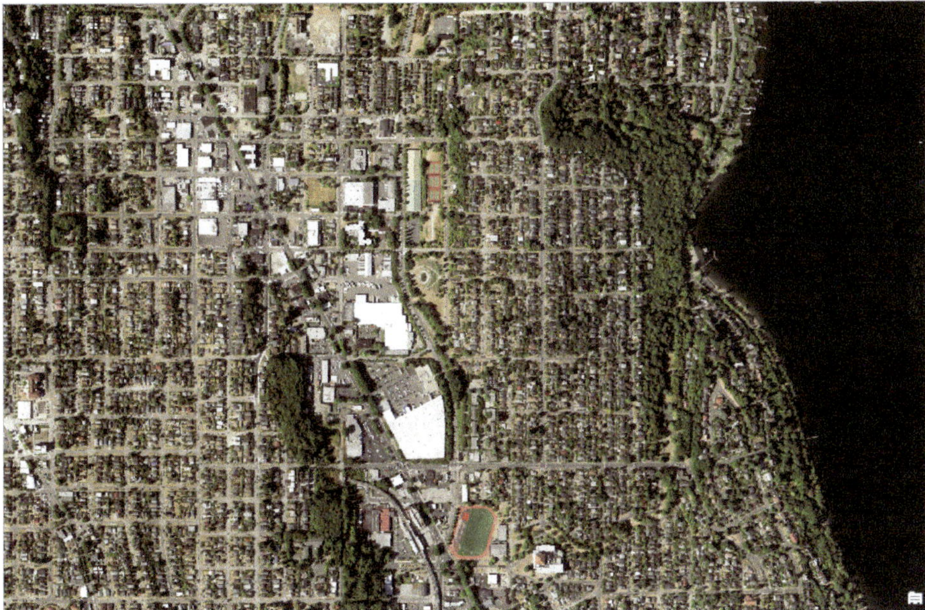

3 In the **Geoprocessing** pane, on the **Environments** tab, under **Processing Extent**, for **Extent**, select the **Current Display Extent** button.

The bounding values for the extent populate. You'll accept the defaults for the other parameters, including for processor type.

4 Click **Run**.

After a few minutes, the process is completed and the Seattle_buildings output layer appears in the Contents pane and on the map. This time, you can see that almost all the buildings were detected.

You've now successfully detected buildings in an area of Seattle using a pre-trained model that has been fine-tuned through transfer learning.

Take the next step (optional)

Now you'll use the Swipe tool to compare the building footprint layers obtained from running the off-the-shelf pretrained model versus the model fine-tuned with transfer learning.

1. From the **Output_provided.gdb** folder, add the polygon feature classes, **Seattle_buildings_off_the_shelf** and **Seattle_buildings_with_transfer_learning**, to see full results that have been already generated.

2. On the ribbon, click the **Feature Layer** contextual tab. In the **Compare** group, click **Swipe**.

3. On the map, click and drag your pointer up and down to see the layer underneath.

   The fine-tuned model does a much better job of identifying the building footprints of smaller buildings in your imagery compared with the off-the-shelf model.

4. Use the **Swipe** tool again to compare the results in the transfer learning layer to the buildings you can observe visually in the imagery.

   You might notice that the layer resulting from the fine-tuned model is still not perfect and a few buildings are missing here and there.

Best practices for fine-tuning models

Fine-tuning a model with transfer learning tends to be an iterative process. You could continue to improve your model's performance by collecting more training examples and conducting another round of transfer learning training. For a quick overview, the steps would be as follows:

First, observe the types of buildings that were missed by the model.

Create new training polygons of these types of buildings and generate new training chips, saving them to a new folder. You should follow the same guidelines as previously, clipping the imagery to ensure that no unlabeled buildings are included in the chips.

Run a new training session, starting with the off-the-shelf pretrained model and feeding it all the chips created so far (that is, for the **Input Training Data** parameter, you'll list all your chip folders). This is best practice to ensure that the model treats all the training chips equally.

Summary

In this tutorial, you used deep learning to extract building footprints from aerial imagery in ArcGIS Pro. You chose a pretrained model from ArcGIS Living Atlas and learned the importance of matching your input data to the model's expectations. You then applied transfer learning to remedy a resolution mismatch; You used a small number of new training samples and further trained the model. You then applied the fine-tuned model to a Seattle neighborhood and obtained enhanced results.

This chapter is based on an Esri tutorial by Sydney Wallace.

CHAPTER 4

Training a SAMLoRA model to identify specific features

Objectives

- Use a small set of training samples to train a deep learning model.
- Extract specific parts of imagery using the trained model.

Introduction

Out-of-the-box pretrained deep learning models may not be effective for specialized needs; in these cases, the SAMLoRA (Segment Anything Model with Low-Rank Adaptation) model may be a good solution. It is based on SAM, a multipurpose foundational model that was trained on massive datasets and uses the LoRA approach to quickly teach the model how to identify specific features after being exposed to only a small number of examples.

In this chapter, you'll train the SAMLoRA model in ArcGIS Pro to identify informal settlements in Alexandra, South Africa. You'll then apply the trained model to extract informal settlement building footprints.

Requirements

- ArcGIS Pro
- ArcGIS Image Analyst
- Deep Learning Libraries for ArcGIS Pro
- Recommended: NVIDIA GPU with a minimum of 4 GB of dedicated memory

Note: For this tutorial, an NVIDIA GPU with a minimum of 4 GB of dedicated memory is recommended. Based on whether your computer has a GPU and what its specifications are, this process may take from under 2 minutes to 20 minutes or more.

If you are not sure whether your computer has a GPU and what its specifications are, see chapter 1.

Tutorial 4-1

Use samples to train a SAMLoRA model

In this tutorial, you'll learn the process of training SAMLoRA to identify features of interest in your imagery—in this case, informal settlement buildings. You'll set up the ArcGIS Pro project, review the examples provided, and export the training data to the format that SAMLoRA expects.

Set up the project

1. In File Explorer, go to **C:/GeoAI_Data** and open the **Chapter04** folder. Double-click **Alexandra_Informal_Settlements.aprx** to open the project in ArcGIS Pro.

   On the map, a drone imagery layer shows a neighborhood in South Africa. The imagery is high resolution, with each pixel representing a square of about 2 × 2 cm on the ground.

2. Explore the built-up areas shown in the image.

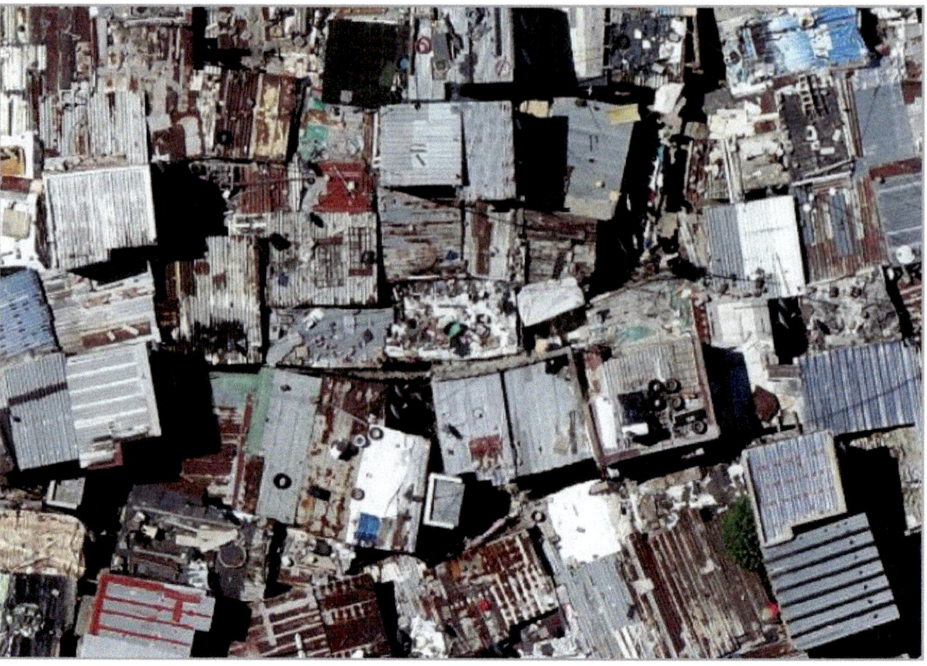

Many of these built-up areas are informal settlements where buildings are built very close to each other, forming intricate patterns. Their roofs are made of corrugated metal sheets of varied colors and maintenance states. For these reasons, traditional deep learning models can have difficulty identifying such buildings with a high degree of accuracy. The SAMLoRA approach that you'll learn in this tutorial is a good alternative to obtain high-quality results without requiring high computing power.

Explore informal settlement examples

You'll review the training examples that were provided with your project.

1. In the **Contents** pane, check the box next to the **Training_Area** layer to turn it on.

   An orange polygon appears on the west side of the imagery. It represents the area chosen to train a SAMLoRA on what informal settlements look like.

2. In the **Contents** pane, right-click the **Training_Area** layer and choose **Zoom To Layer**.

   The map zooms in to the training area.

3 Turn on the **Informal_Settlements_Examples** layer.

The Informal_Settlements_Examples layer represents all the buildings in the training area as light-gray polygons. Every building in the training area was captured as a polygon. If even a few buildings were missing, it would generate confusing information and SAMLoRA would not train optimally. In the following example images, on the left, the training set is complete and ready to be used for training. In contrast, on the right, some buildings are missing, which would yield poor training performance.

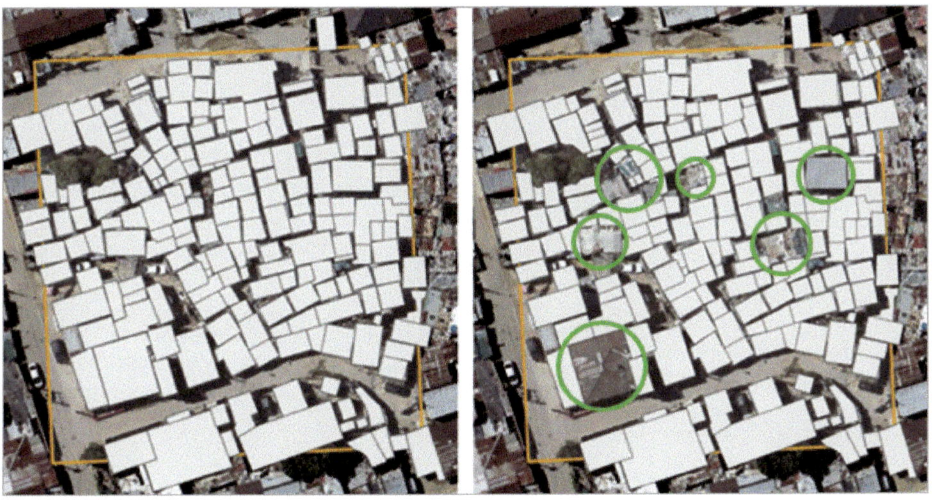

Next, you'll inspect the Informal_Settlements_Examples attribute fields.

4 In the **Contents** pane, right-click the **Informal_Settlements_Examples** layer and click **Attribute Table**.

Every line in the table represents one of the building polygons. The Class attribute has the value Building.

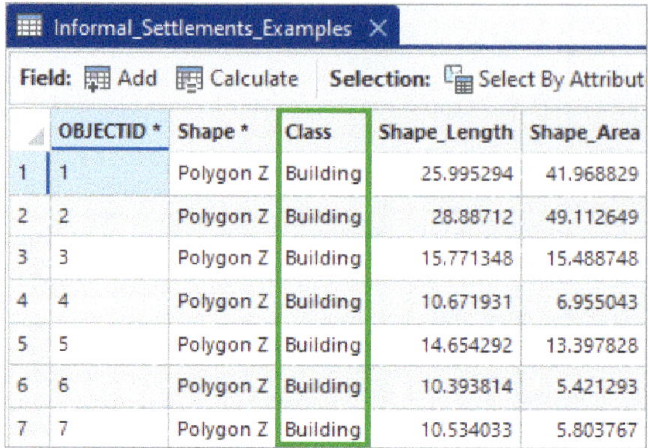

Informal_Settlements_Examples

Field: Add Calculate Selection: Select By Attribut

	OBJECTID *	Shape *	Class	Shape_Length	Shape_Area
1	1	Polygon Z	Building	25.995294	41.968829
2	2	Polygon Z	Building	28.88712	49.112649
3	3	Polygon Z	Building	15.771348	15.488748
4	4	Polygon Z	Building	10.671931	6.955043
5	5	Polygon Z	Building	14.654292	13.397828
6	6	Polygon Z	Building	10.393814	5.421293
7	7	Polygon Z	Building	10.534033	5.803767

Note: Although this is not an essential part of this workflow, note that the value **Building** is a label for the underlying numeric value **1.** In a case in which the training examples represent several feature types, the **Class** field would list these several types—for instance, **Building, Road,** or **Tree.** This approach enables SAMLoRA to learn how to recognize different feature types in your imagery.

5 Close the table.

6 In the **Contents** pane, uncheck the box next to **Informal_Settlements_Examples** to turn the layer off and reduce clutter on the map.

Learn about training chips and cell size

You'll use the example polygon layer and the imagery layer to generate training data in a specific format. When producing training chips, one important thing to determine is their optimum size. To detect objects that are close to each other, as is the case with informal settlement buildings, a good guideline is that a chip should include 6 to 12 features. Another way to think about it is that a chip should contain at least one or two complete features in its center and show a good amount of context (or background) around it.

The following example image contains three chips of different sizes:

- Chip 1 is too small; it contains no complete features and almost no context.
- Chip 2 is a good size; there are two complete features in the center, as well as a few incomplete features and context around them.
- Chip 3 is too large; it contains many features (more than 25).

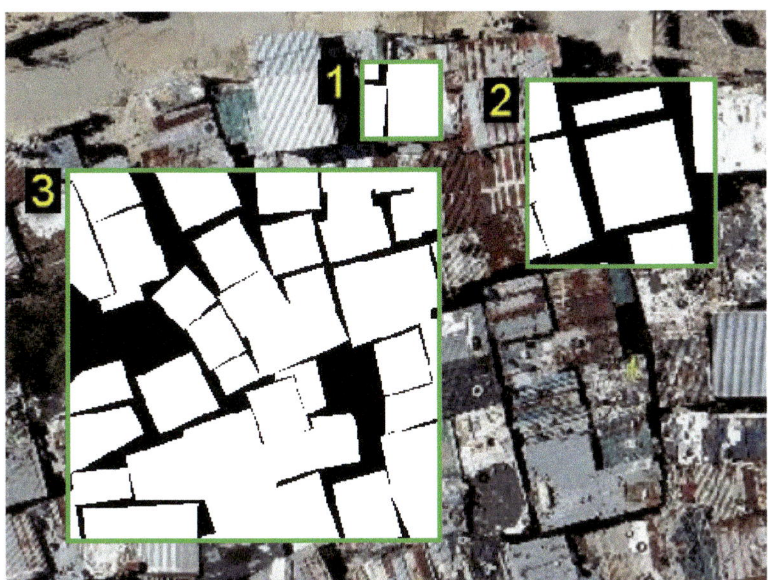

It is customary to have chips that measure 256 × 256 pixels (or cells), so you'll keep that default value. However, you can obtain chips of different sizes by varying the size of the cells that compose them. The optimum chip size will depend on the size of the features you want to identify.

Export training data

Next, you'll generate the training chips using the Export Training Data For Deep Learning tool.

1. On the ribbon, click the **Analysis** tab. In the **Geoprocessing** group, click **Tools**.
2. In the **Geoprocessing** pane, search for and open the **Export Training Data For Deep Learning (Image Analyst Tools)** tool. Apply the following settings:
   - For **Input Raster**, select **Alexandra_Orthomosaic**.
   - For **Output Folder**, type Informal_Settlements_256_5cm. (This creates a subfolder in your project to hold the training data.)
   - For **Input Feature Class Or Classified Raster Or Table**, select **Informal_Settlements_Examples**.
   - For **Class Value Field**, select **Class**.
   - For **Input Mask Polygons**, select **Training_Area**.

     Specifying the Training_Area layer as a mask is crucial because it will ensure that the image chips will be created only within the area where all the buildings are labeled.

   - For **Metadata Format**, select **Classified Tiles**.

     Classified Tiles is the metadata format that is expected to train a SAMLoRA model.

3. In the tool pane, click the **Environments** tab. For **Cell Size**, type 0.05.

   This cell size is 0.05 meters, or 5 centimeters. As indicated in the previous section, this cell size will ensure that the tiles generated are of a suitable size to identify informal settlement buildings.

   Note: Choosing the cell size for your image chips will not change the original resolution of your input imagery. The tool will resample the data on the fly to produce image chips of the desired cell size.

4. Click **Run**.

   The SAMLoRA deep learning model will use the labeled tiles to learn how the buildings look and where they are located. You can examine the training in the Catalog pane.

Train a SAMLoRA informal settlement model

Next, you'll use the training data you generated to train the SAMLoRA deep learning foundational model and teach it to identify the informal settlement buildings in your imagery. Then, you'll review the trained model to better understand it.

1. Search for and open the **Train Deep Learning Model** tool and apply the following settings:

    - For **Input Training Data**, click the browse button. Browse to **Folders** > **Alexandra_Informal_Settlements**, select **Informal_Settlements_256_5cm**, and click **OK**.
    - For **Output Folder**, type Informal_Settlements_256_5cm_SAMLoRA.

      This parameter creates a subfolder in your project (under Folders > models) to hold the resulting trained model.

    - For **Max Epochs**, type 50.

      An epoch refers to one complete pass of the entire training dataset.

    - For **Model Type**, select **SAMLoRA (Pixel classification)**.

2. Expand **Data Preparation**. For **Batch Size**, if your NVIDIA GPU dedicated memory is 4 GB, type 2. If it is 8 GB, type 4.

> **Note:** You can examine your GPU memory usage live by opening the command prompt from the Windows start menu and pasting the command `nvidia-smi -l -5`.
>
> 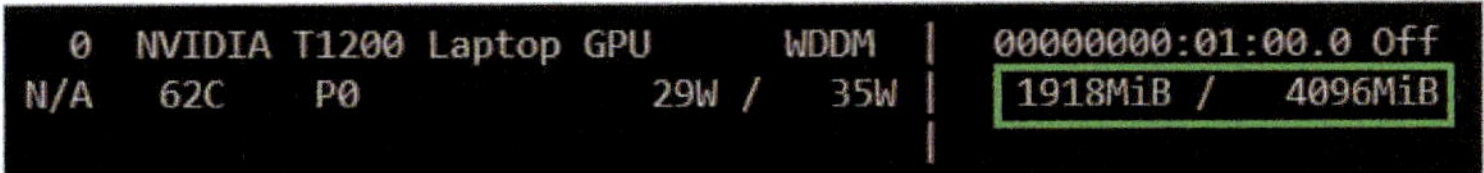
>
>
> If you aren't reaching the maximum, you can increase the batch size the next time you run the tool.

3. Expand **Advanced** and apply the following settings:

    - For **Backbone Model**, select **ViT-B**.

      B stands for basic. This option trains the SAMLoRA model on a smaller (basic) neural network.

- For **Monitor Metric**, confirm that **Validation loss** is selected.

  This metric measures how well the model generalizes what it has learned to new data.

- Confirm the **Stop when model stops improving** option is checked to avoid model overfitting.

  You are now ready to run the tool.

4. Click **Run**.

5. Click **View Details**, and in the details window, click the **Messages** tab and monitor the **Validation loss** metric.

   The smaller the validation loss value, the better. It gradually decreases with each epoch as the model improves its ability to successfully identify building areas. When its value no longer changes significantly, the training stops. In parallel, the accuracy and Dice metrics (third and fourth columns) steadily increase. These metrics measure the model's performance.

6. When the training ends, review the **accuracy** and **Building precision** numbers.

```
15         0.161921426653862    0.15452569723129272  0.8786553144454956   0.9225639700889
{'accuracy': '8.9250e-01'}
           NoData   Building
precision  0.896398  0.891348
recall     0.709159  0.966789
f1         0.791860  0.927537
```

   The values you obtain might be different. In the example image, the overall accuracy is 8.9250e-01, or 89.25 percent. For this use case, it should be between 85 and 95 percent. Because you are specifically interested in identifying buildings, the precision value for Building is the best measure of the model's performance. In this case, it is 0.8913 or 89.13 percent.

7. Close the details window.

Review the trained SAMLoRA model

One way to learn more about your trained SAMLoRA model is to look at it through the Review Deep Learning Models tool.

1. On the ribbon, click the **Imagery** tab. In the **Image Classification** group, click **Deep Learning Tools** and select **Review Deep Learning Models**.
2. In the **Deep Learning Model Reviewer** tool, for **Model**, click the **browse** button.
3. Browse to **Folders** > **Alexandra_Informal_Settlements** > **models**, select **Informal_Settlements_256_5cm_SAMLoRA**, and click **OK**.
4. Locate the graph under **Training and Validation Loss**.

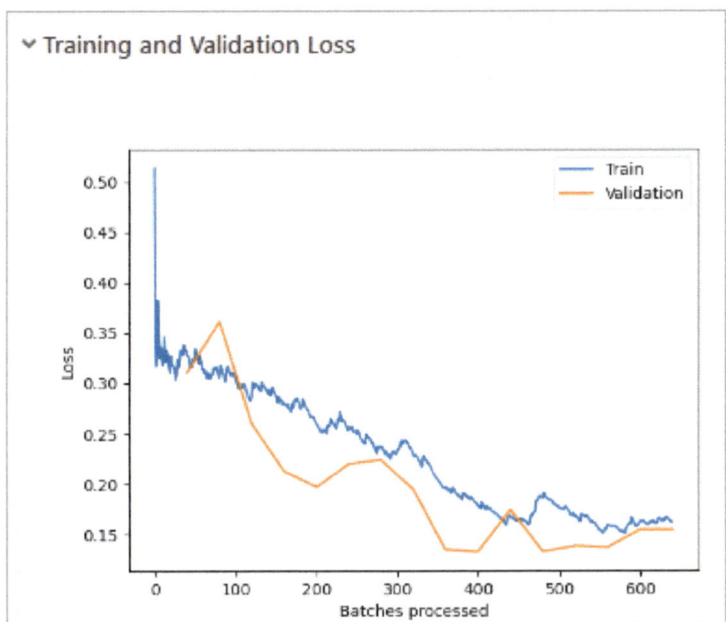

This graph shows the detail of how the model learned. Training Loss (in blue) shows how well the model learned on the training data, and Validation Loss (in orange) shows how the model was able to generalize what it had learned to new data. (Your graph might look different.) In the last part of the training, the training and validation loss curves were forming asymptotic lines.

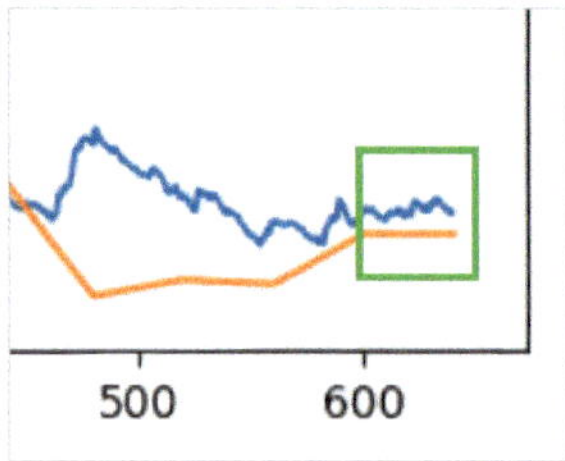

This phenomenon is called convergence. If the training continues beyond that phase, the model might start performing better on training data than on validation data, a sign that it is overfitting the training data and losing its ability to generalize to new data.

5. Review the other information provided in the pane, including **Model Type**, **Backbone**, overall **Accuracy**, and **Epochs Details**.

6. When you're finished reviewing the model information, close the **Deep Learning Model Reviewer** pane.

You have now trained the SAMLoRA foundational deep learning model to identify informal settlement buildings.

Tutorial 4-2

Classify informal settlements with the trained SAMLoRA model

You are now ready to run your model and extract informal settlements from the imagery. You'll apply the trained SAMLoRA model to your imagery to classify pixels as building or no building. You'll then derive building footprint polygons. Finally, you'll explore expanded results to understand the power of the SAMLoRA approach to extract different object types for larger extents.

Classify informal settlements

First, you'll apply the trained SAMLoRA model to your imagery using the Classify Pixels Using Deep Learning tool. To expedite the workflow in this tutorial, you'll run the process on only a small extent. However, in real life, you could process large amounts of imagery.

1. On the ribbon, click the **Map** tab. In the **Navigate** group, click **Bookmarks** and click the **Inferencing Area** bookmark.

   The map zooms to the bookmark.

2. Search for and open the **Classify Pixels Using Deep Learning** tool and apply the following settings:

   - For **Input Raster**, select **Alexandra_Orthomosaic.**
   - For **Output Raster Dataset**, type Informal_Settlements_Raster.
   - For **Model Definition**, click the browse button. Browse to **Folders** > **Alexandra_Informal_Settlements** > **models** > **Informal_Settlements_256_5cm_SAMLoRA**, select **Informal_Settlements_256_5cm_SAMLoRA.dlpk**, and click **OK**.
   - For **Batch Size**, type the same value you used previously (for instance, 2 if your GPU dedicated memory is 4 GB).

3. In the tool pane, click the **Environments** tab. Under **Processing Extent**, click the **Current Display Extent** button.

   This parameter ensures only the area currently displaying on the map will be processed.

4. For **Cell Size**, type 0.05 to match the cell size you used when training the SAMLoRA.

   This tool process can take from under a minute to up to 15 minutes, depending on whether you have a GPU and what its specifications are.

5. Click **Run**.

   When the process is complete, the new raster layer appears on the map.

Every pixel in the imagery for your chosen extent was classified, and the result was captured in the output raster: Areas pertaining to an informal settlement building were assigned a value of 1—symbolized in light orange—and non-building areas were assigned a value of 0—symbolized as transparent.

The model could be applied to a much larger imagery extent, covering an entire city or region.

Take the next step (optional)

Now that you've generated an output raster, you could perform some post-processing on it to derive a building footprints layer. To do this, you could use the standard tools for this process or use a custom tool that includes the following main steps:

1. Use the **Raster to Polygon** tool to obtain a polygon layer.
2. Use the **Regularize Building Footprints** tool to normalize and smooth out the polygon shapes.
3. Remove the polygons that are too small to be buildings.
4. Use the **Dissolve** tool to remove overlapping polygons.

Best practices for using the SAMLoRA approach

To apply this workflow to your own imagery, keep the following tips in mind:

- Storing imagery: When working with your own data, you can similarly host it in ArcGIS Online. See the Publish Hosted Imagery Layers documentation page at link.esri.com/ExploreGeoAI/HostedImagery to learn more. Another option is to use imagery that is stored on your local computer.
- Using consistent imagery throughout the workflow: When working with the SAMLoRA model, you should ensure that you are using similar imagery to both train and apply the model. Most particularly, the spectral bands (such as red, green, and blue), pixel depth (such as 8 bit), and cell size should be the same.
- Finding your imagery properties: If you are not certain what your imagery properties are, in the **Contents** pane, right-click your imagery layer and choose **Properties**. In the **Properties** pane, click the **Source** pane. Under **Raster Information**, find the **Number of Bands**, **Cell Size X**, **Cell Size Y**, and **Pixel Depth** values.
- Using an attribute domain: In the training examples polygon layer, the value **Building** in the **Class** attribute is a label for the underlying numeric value 1. Although not an essential part of the workflow, you should know that this is implemented with an attribute domain. Alternatively, you could use numeric values for your classes; the output will also be numeric, the extracted features being **1** instead of **Building**.
- Creating a training area layer: You can use the **Create Feature Class** geoprocessing tool. Then, on the ribbon, on the **Edit** tab, click the **Create** tool to trace one or several rectangle polygons outlining the areas where you want to provide training examples.
- Choosing the cell size: Remember to experiment with the cell size to generate tiles that are optimized for the features you plan to extract.
- Experimenting on a small extent: While experimenting, you can limit the processing to a small extent for faster results. On the **Environments** tab, under **Processing Extent**, click the **Draw Extent** button and draw a small polygon on the map. Alternatively, zoom in on the map and click the **Current Display Extent** button.

Summary

In this tutorial, you used the SAMLoRA approach to identify informal settlements in your imagery. You generated training data and used it to train the foundational model. You applied the trained model to classify informal settlements in your imagery and then derived and cleaned up building footprint polygons. Finally, you explored expanded results.

This chapter is based on an Esri tutorial by Rami Alouta, Colin Kelly, and Delphine Khanna.

CHAPTER 5

Classifying objects using deep learning

Objectives

- Prepare data for use in training a deep learning model in ArcGIS Pro.
- Use a training sample to train a deep learning model in ArcGIS Pro.
- Apply an object classification model to a given scenario in ArcGIS Pro.

Introduction

The 2018 Woolsey Fire burned thousands of acres in Southern California. In the past, analysts had to manually review aerial images to determine the scope of building damage. Deep learning can automate this process.

In this chapter, you will play the role of a wildfire analyst. You will prepare aerial imagery and training sample data. Then, you will train a deep learning model to classify buildings as damaged or undamaged and evaluate the model's accuracy. Using the trained model, you will assess the damage of the buildings in the aerial images.

Requirements

- ArcGIS Pro
- ArcGIS Image Analyst
- Deep Learning Libraries for ArcGIS Pro
- Recommended: NVIDIA GPU with a minimum of 4 GB of dedicated memory

Note: For this tutorial, an NVIDIA GPU with a minimum of 4 GB of dedicated memory is recommended. Based on whether your computer has a GPU and what its specifications are, this process may take from under 2 minutes to 20 minutes or more.

If you are not sure whether your computer has a GPU and what its specifications are, see chapter 1.

Tutorial 5-1

Prepare and train a deep learning classification tool

Set up the project

First, you will open the ArcGIS Pro project that contains the training samples of aerial imagery from the Woolsey wildfire and Los Angeles County building footprints.

1. In File Explorer, open the **C:/GeoAI_Data** folder. In the **Chapter05** folder, double-click **Woolsey Fire.aprx** to open it in ArcGIS Pro.

The project opens to the area where the Woolsey wildfire occurred. In the Contents pane, the Training Samples layer contains polygons of building footprints that have already been classified as damaged (in red) or undamaged (in green). A large input dataset is necessary to train the model to detect structures and classify them properly.

Before reviewing the buildings on the map, you will explore the two building class attributes and how they are stored in the Training Samples feature class.

Review the training data

You'll review the building footprints that have been classified as damaged or undamaged to ensure that they are accurate for training a deep learning model. For the purposes of this exercise, the building footprints have been provided.

1 In the **Contents** pane, right-click the **Training Samples** layer and click **Data Design** > **Fields**.

2 In the **Fields** view, in the **Domain** column, locate the **Class_Value** field.

The Class_Value field is used to classify the buildings as Damaged or Undamaged. It has a domain named class_value that controls the values that can be entered for this attribute field.

You will explore how the domain is storing these values.

3 On the ribbon, click the **Fields** tab. In the **Data Design** group, click **Domains**.

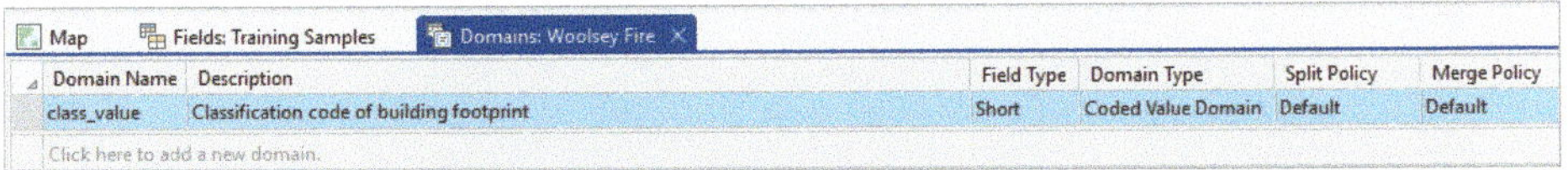

The Domains view appears. It displays all the domains for this project's geodatabase. There is only the one domain that you saw earlier (class_value), which is already selected. Under Description, the values of Damaged and Undamaged are listed.

In the domains view, in the Code column, notice that the Code value is stored in the geodatabase as 1 or 2.

The deep learning model that you will create needs to use numeric values as inputs for the classes. This domain allows you to see readable values, such as Damaged or Undamaged, while feeding values of 1 and 2 into your training data for the deep learning tools to process.

4 Close both the **Domains** view and the **Fields** view.

Now that you understand how the values are stored, you will ensure that the Training Samples polygons are properly classified.

5 On the ribbon, click the **Edit** tab. In the **Selection** group, click **Attributes**.

The Attributes pane appears. You will review features in this layer and their attributes.

6 In the **Attributes** pane, click the **Layers** tab.

7 Click **Choose A Layer** to open the list and select **Training Samples**.

A list of features appears, and the first building is selected; its attributes are shown below the list. You will go through some of the building features and confirm that the Class Value attribute reflects each building's status.

8 In the **Attributes** pane, click the **Step Forward** button to progress through the buildings in this layer.

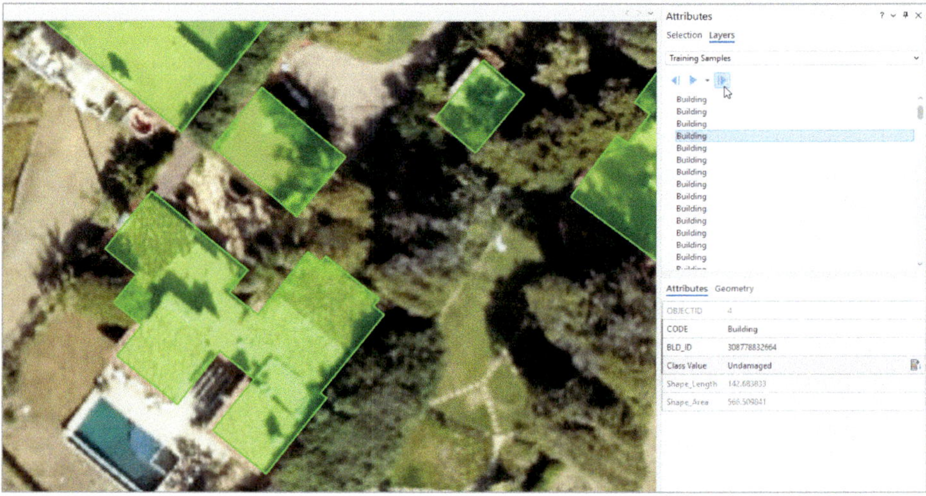

Errors in the training data will reduce its overall accuracy. In the Attributes pane, you can update the Class Value attribute.

9 Close the **Attributes** pane when you have familiarized yourself with the workflow.

You have finished reviewing the building footprints that have been classified as damaged or undamaged by the Woolsey wildfire to ensure that they are accurate for training a deep learning model.

Export training samples

A deep learning model cannot process the raw aerial imagery and the Training Samples polygons. Instead, you will split the image into smaller tiles, called chips, and convert the polygons to labels. The chips are smaller pieces of imagery for the software to process, and the labels will match the different classes of features that you identified as either damaged or undamaged. In this step, you will use the Export Training Data For Deep Learning geoprocessing tool to perform this action.

1 On the ribbon, click the **Analysis** tab. In the **Geoprocessing** group, click **Tools**.

2 In the **Geoprocessing** pane, search for and open the **Export Training Data For Deep Learning (Image Analyst Tools)** tool and apply the following settings:

- For **Input Raster**, select **USAA Imagery (11/13/18)**.
- For **Output Folder**, type buildings_training_samples.
- For **Input Feature Class Or Classified Raster Or Table**, select **Training Samples**.
- For **Class Value Field**, select **Class Value**.
- For **Tile Size X**, type 448.
- For **Tile Size Y**, type 448.
- For **Metadata Format**, select **Labeled Tiles**.

3 In the **Geoprocessing** pane, click the **Environments** tab. Under **Raster Analysis**, for **Cell Size**, type 0.3.

A deep learning model performs best when the training and prediction images are the same resolution. Cell size is the dimensions of the image pixels in real-world units. For instance, if you are using a projected coordinate system with linear units of meters, setting the cell size to 0.3 makes each pixel in the output 0.3 meters wide.

 Click **Run**.

The Export Training Data For Deep Learning tool may take several minutes to run. Once finished, you successfully exported a set of image tiles with labels that identify the buildings as damaged or undamaged.

Train the feature classifier model

Using the tiles that you created, you will train a deep learning model. You will use the chips and labels that you exported to create a deep learning model for classifying buildings as damaged or undamaged.

Depending on your computer's hardware, training the model can take 30 minutes or more.

1 Search for and open the **Train Deep Learning Model** tool and apply the following settings:

- For **Input Training Data**, click the browse button. On the **Input Training Data** dialog box, browse to **GeoAI_Data\Chapter05**, select **buildings_training_samples**, and click **OK**.
- For **Output Folder**, type buildings_classify.

  Now, you will set the number of epochs for your model to train. The more times that you pass the data through this network, the better your results will be—although after a while, the benefit from doing so will decrease.
- Confirm that **Max Epochs** is set to **20**.

  Next, you will choose the type of model to produce, which is based on the format of your training data from the previous exercise.
- Confirm that **Model Type** is set to **Feature classifier (Object classification)**.

  Next, you will set the Batch Size value. This parameter is the number of training samples that will be processed at a time.

2 Expand the **Data Preparation** section. For **Batch Size**, type 8.

3 If your computer has a GPU, near the top of the **Geoprocessing** pane, click the **Environments** tab, and then for **Processor Type**, select **GPU**.

4. Click **Run**.

> **Note:** If the model fails to run, you may have to adjust the **Batch Size** parameter. The ability to process larger batch sizes is based on your computer's GPU. If you have a powerful GPU, you can process up to 64 training samples at a time. If you have a less powerful GPU, you may have to set this parameter to 8, 4, or 2 and rerun the tool.
>
> The time that it takes to finish will depend on your computer's hardware. The training will stop automatically when the accuracy of the model stops improving or when the number of epochs reaches the specified limit.

5. Near the bottom of the **Geoprocessing** pane, click **View Details**.
6. Click **Messages**.

   For each of the epochs that run, Messages displays the model's performance. Specifically, you will begin to see the accuracy of the model increase or move toward 1.

7. After the tool finishes running, close the details window.

   You have trained a deep learning model for classifying buildings as damaged or undamaged.

Review the results of training your model

In this step, you will observe the results from training the model that you created in the previous section.

1. In File Explorer, browse to **C:\GeoAI_Data\Chapter05\models** and open the **buildings_classify** folder.

   The folder contains several outputs that you can use to process imagery:

   - **buildings_classify.pth** is a trained model and is usually saved in a PyTorch format.
   - **buildings_classify.emd** is a model definition file that contains model information about the tile size, classes, and model type.
   - **model_metrics.html** contains details about the learning rate used and the final accuracy of the model.

- **buildings_classify.dlpk** is a complete package of all the files stored in the model output folder, including the trained model, the model definition file, and the model metrics file. This package can be shared to ArcGIS Online and ArcGIS Enterprise as an item for others to access.

Next, you will explore the model_metrics.html page to see how your model performed during training. Because the model trains differently each time the tool is run, your model's metrics may vary.

2 Double-click the **model_metrics.html** file.

Your default web browser opens.

3 Locate the **Training and Validation Loss** graph.

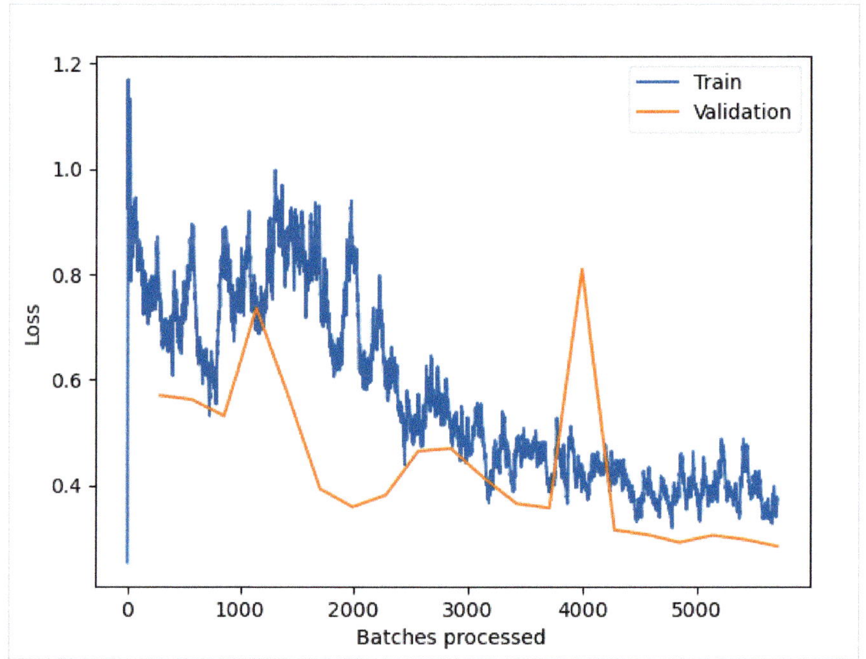

The Training and Validation Loss section displays a graph of the amount of error that was present as the model trained over time. When the model ran, some of your input data was used to train the model, and some was used to validate the model or test the model to determine its accuracy. Ideally, you would see these values decrease as the number of images processed increases over the course of the 20 epochs.

4 Under **Analysis of the model**, locate the **Confusion matrix**.

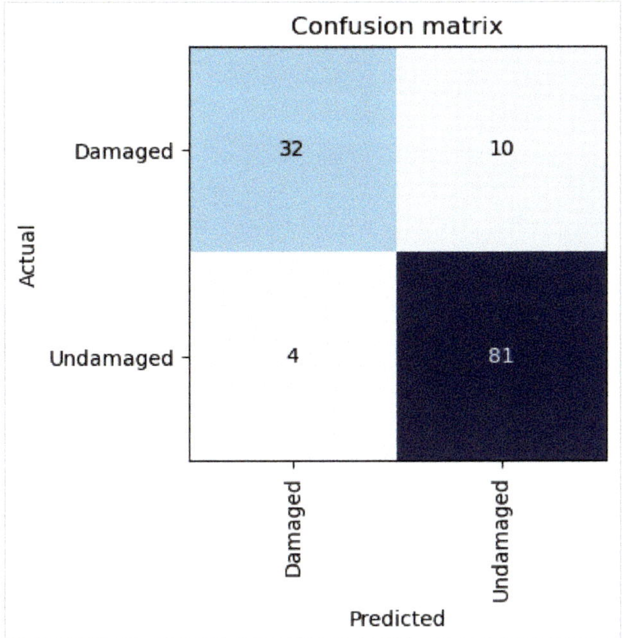

The Confusion Matrix graph visualizes how well the model identified buildings between the two classes using the actual, or input, data and the predicted results.

5. Under **Sample Results**, locate the **Ground truth/Predictions** section.

The Ground truth/Predictions section provides some examples of the training's results. Images are shown side by side, with the known results on the left and the predictions on the right.

6. Close the **model_metrics.html** web browser tab.

You have now trained a deep learning model and examined its metrics.

Tutorial 5-2

Use the trained model to classify features

With the model trained, you will use it to help expedite identifying, or inferencing, damaged and undamaged buildings. Then, you will observe the results.

Prepare the imagery

First, you will use your model on a subset of the imagery provided to you. You will look at a part of the map that has not yet been classified.

1. In the **Contents** pane, turn off the **Training Samples** layer and turn on the **Building Features** layer.

2. Right-click the **Building Features** layer and click **Zoom To Layer**.

   The Building Features layer has additional building footprints in the southeast part of the map that need to be classified.

3. On the **Map** tab, in the **Navigate** group, click the **Bookmarks** button and then click the **Inference Area** bookmark.

You will use the trained model to classify the buildings in this extent.

Configure the deep learning tool

The Classify Objects Using Deep Learning tool uses a trained model to make classification predictions. The input raster gets split into tiles, which are fed in batches to the model to classify features. The classifications, after some postprocessing, are written into a specified output feature class.

1. Search for and open the **Classify Objects Using Deep Learning** tool. Apply the following settings:

    - For **Input Raster**, select **USAA Imagery (11/13/18)**.
    - For **Input Features**, select **Building Features**.
    - For **Output Classified Objects Feature Class**, type Inferenced_Buildings.
    - For **Model Definition**, click the browse button. On the **Model Definition** dialog box, browse to **C:\GeoAI_Data\Chapter05\model\buildings_classify**, select the **buildings_classify.dlpk** deep learning model package file, and click **OK**.
    - For **Class Label Field**, leave the default value of **ClassLabel**.
    - Under **Arguments**, for **Batch Size**, type 8.
    - Leave the **Test Time Augmentation** value as **False**.

    Before running the tool, you will set some environments.

2. In the **Geoprocessing** pane, click the **Environments** tab. Apply the following settings:

    - Under **Processing Extent**, for **Extent**, click the **Current Display Extent** button.

    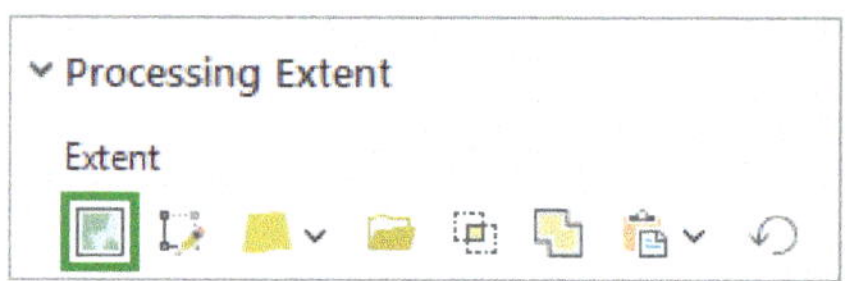

    After you choose Current Display Extent, the coordinates of the extent's geographic bounding box are displayed.

    - Under **Raster Analysis**, for **Cell Size**, type 0.3.
    - Under **Processor Type**, set **Processor Type** to **GPU**.

3. Click **Run**.

> **Note:** This tool can take 30 minutes or more to run, depending on your computer's hardware. The ability to set the **Batch Size** to 8 depends on your GPU. If the tool fails to run, try changing this parameter to 4 or 2.

The results are added to the map.

4. In the **Contents** pane, turn off the **Building Features** layer.

The results of running the Classify Objects Using Deep Learning tool are now added to your map.

Review the results

Now that your results have been added to the map, you will see how accurate they are. To help with this process, you will use a layer file that contains appropriate symbology to update the damaged and undamaged features so that they are displayed in different colors.

1. In the **Contents** pane, click the **Inferenced_Buildings** layer.

2. On the ribbon, click the **Feature Layer** contextual tab. In the **Drawing** group, click the **Import** button.

   The Import Symbology dialog box appears.

3. On the **Import Symbology** dialog box, for **Symbology Layer**, click the **browse** button. Browse to **C:\GeoAI_Data\Chapter05** and select the **Inferenced Buildings Symbology.lyrx** file. Click **OK**.

   Notice that the Import Symbology tool's parameters are automatically populated.

4. Click **OK**.

The Inferenced_Buildings layer now shows the damaged buildings in red and the undamaged buildings in green.

Next, you will use the Attributes pane to review your model's results.

5. On the **Edit** tab, in the **Selection** group, click the **Attributes** button.
6. In the **Attributes** pane, click the **Layers** tab, if necessary.
7. Change the current layer to **Inferenced_Buildings**.
8. Click the **Play** button to start progressing through the building footprints.

> **Note:** You can change the play speed by clicking the arrow next to the **Play** button and choosing a different speed.

9. In the map, observe how well your model classified the buildings as damaged or undamaged.

> **Note:** If you want to change the classification for a feature, click the **Pause** button and update the **ClassLabel** attribute and save any edits.

10. Save your project.

Summary

In this chapter, you prepared data to train a deep learning model, trained the model, classified a set of features using your model, and examined the results.

Being able to train a model to automatically classify features can save an organization valuable time and money while reducing the possibility of human error, which is especially critical when time is limited and people's lives and property are at risk.

This chapter is based on an Esri tutorial by David Yu.

CHAPTER 6

Using deep learning for point cloud classification

Objectives

- Use LAS format point clouds to train a classification model.
- Determine the best epoch after training the model.
- Classify points into two classes to conduct a risk assessment analysis.

Introduction

You can use deep learning to classify LAS format point clouds and classify many kinds of features. It doesn't use predefined rules to identify specific things such as buildings or ground. Rather, you provide examples of features of interest, and those are used to train a neural network that can then recognize and classify those features in other data.

In this chapter, you'll use a deep learning model to conduct a risk assessment analysis of power lines using lidar data. You'll analyze an area in Northern California that has potential fire risk due to trees near power lines. Conducting a risk assessment is a technique used by insurance companies for evaluating and assessing the risks associated with an insurance policy, which helps determine the applicable premium for the customer. Lidar is an optical remote sensing technique that uses laser light to provide elevation information, which can better model the spatial relationship between power lines and the surrounding area to aid in the assessment. You'll use ArcGIS Pro to train a deep learning model to identify the power lines from a lidar point cloud. Deep learning allows you to train a model using a sample dataset and apply the model to other similar areas. You'll assess which model resulted in the most accurate results and use it to classify the lidar points that are power lines.

Requirements

- ArcGIS Pro
- ArcGIS 3D Analyst™
- Microsoft Excel
- Deep Learning Libraries for ArcGIS Pro
- Recommended: NVIDIA GPU with a minimum of 4 GB of dedicated memory

Note: For this tutorial, an NVIDIA GPU with a minimum of 4 GB of dedicated memory is recommended. Based on whether your computer has a GPU and what its specifications are, this process may take from under 2 minutes to 20 minutes or more.

If you are not sure whether your computer has a GPU and what its specifications are, see chapter 1.

Tutorial 6-1

Train point cloud classification data

Set up the project

You'll open an ArcGIS Pro project and view the training data and validation data.

1. In File Explorer, navigate to the **C:\GeoAI_Data** folder. Open the **Chapter06** folder.

   All the data required to complete the tutorial is in three folders:

   - **smalldata**: Contains a small training dataset and the training boundary, a small validation dataset and the validation boundary, and a DEM raster.
   - **testdata**: Contains a test dataset, the processing boundary, and a DEM raster.

- **results**: Contains the training results from the large dataset that you can use instead of training the model for the large dataset. In the final section of this tutorial, you'll use this trained model to classify a LAS dataset.

Some of the main files in the exercise dataset are LAS files. A LAS file is an industry-standard binary format for storing lidar data. The LAS dataset allows you to examine LAS files, in their native format, quickly and easily, providing detailed statistics and area coverage of the lidar data contained in the LAS files.

2 Double-click **Powerline Classification with DL.aprx** to open it in ArcGIS Pro.

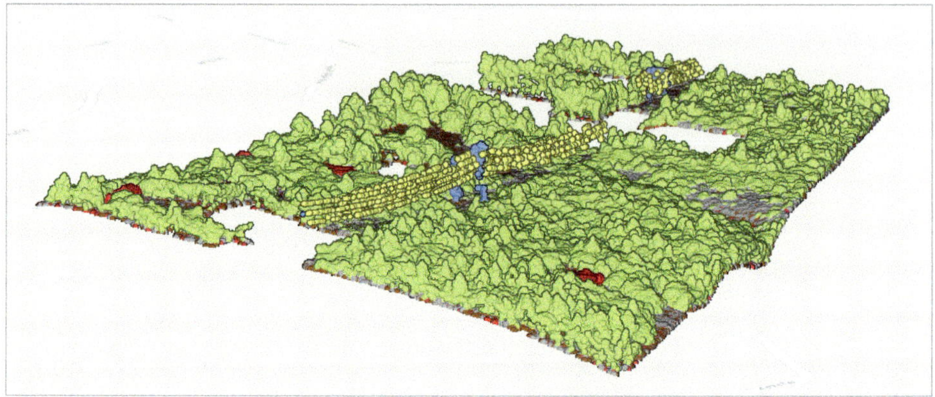

In the map, the val_small.lasd and train_small.lasd layers are symbolized by class codes.

3 Toggle visibility for the layers on and off and zoom in and out to familiarize yourself with the project. When you are ready to continue, turn both layers on.

View the training data and validation data

First, you'll explore these class codes.

1 In the **Catalog** pane, expand **Folders** > **Chapter06** > **smalldata**. Right-click **train_small.lasd** and click **Properties**.

2 In the **LAS Dataset Properties** window, click the **Statistics** tab.

The dataset contains several class codes:

- **1 Unassigned** represents unclassified points, mainly low vegetation or objects above the ground but lower than high vegetation and buildings.

- **2 Ground** represents the ground.
- **5 High Vegetation** represents high vegetation, such as trees.
- **6 Building** represents buildings and other structures.
- 7 **Noise** represents low points.
- **14 Wire - Conductor** represents actual power wires.
- **15 Transmission Tower** represents the power line towers.

You are interested in class code 14 Wire - Conductor. Your goal is to train a model to detect lidar points that are the power lines.

There is a message that states that one file has out-of-date or no statistics. You'll fix this issue now so that you can display the correct class codes in the scene.

3 Click **Update**.

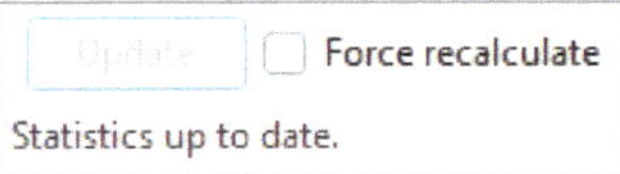

4 Click **OK**.

Next, you'll train a model using these two datasets that you explored.

Prepare the training dataset

LAS datasets cannot be used directly to train the model. The LAS datasets must be converted into smaller training blocks. You'll use the Prepare Point Cloud Training Data geoprocessing tool in ArcGIS Pro to export the LAS files to blocks. You'll use the small datasets to train the model.

> **Note:** The small dataset contains a limited number of points, which reduces the file size and training time. It is designed to give you a good understanding of the process, but it is not intended to train a highly accurate model.

Your goal is to train the model to identify and classify the points that are power lines. Only the points near the power lines are of interest. You'll use a boundary layer, bnd_train_small and bnd_val_small, to specify areas where points should be converted into training blocks.

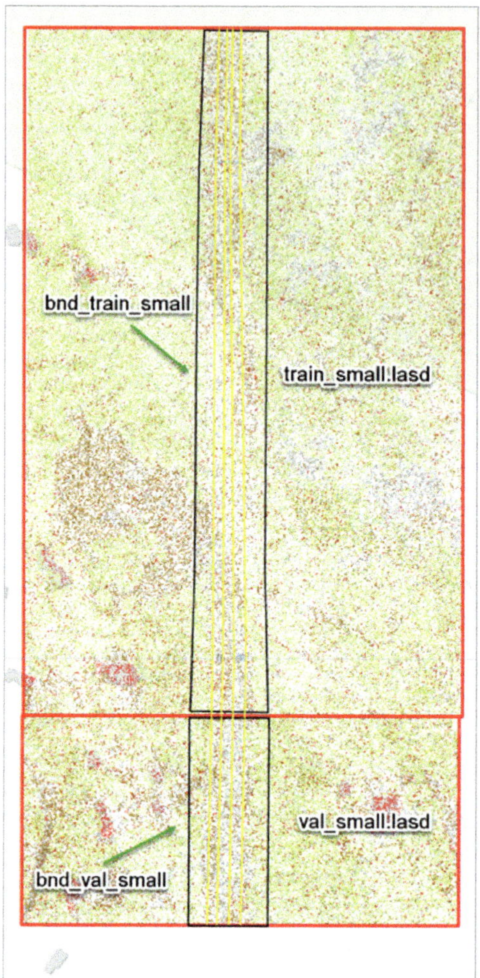

1. In the **Geoprocessing** pane, search for and open the **Prepare Point Cloud Training Data** tool.

   You have entered all the training data, validation data, and processing boundaries.

2. Apply the following settings:

   - For **Input Point Cloud**, select **train_small.lasd**.
   - For **Training Boundary Features**, click the browse button. Browse to **C:\GeoAI\Chapter06\smalldata\boundaries.gdb**, select **bnd_train_small**, and click **OK**.
   - For **Validation Point Cloud**, select **val_small.lasd**.

- For **Validation Boundary Features**, browse to **C:\GeoAI\Chapter06\smalldata\boundaries.gdb**, select **bnd_val_small**, and click **OK**.
- For **Reference Surface**, browse to **C:\GeoAI\Chapter06\smalldata**, select **dem_small.tif**, and click **OK**.

The DEM raster will be used to calculate points' relative height from the ground.

Next, you'll specify which class codes to exclude.

3 For **Excluded Class Codes**, type 2. Click **Add another** and type 7.

Ground (class 2) and noise (class 7) points are excluded from the training data. Ground points typically account for a large portion of the total points, so excluding ground points will make the training process go more quickly.

4 Leave **Filter Blocks by Class Code** blank.

5 For **Output Training Data**, browse to **C:\GeoAI\Chapter06\results**. For **Name**, type training_data_small.pctd. Click **Save**.

Note: The output file extension **.pctd** stands for point cloud training data.

6 Continue entering the following parameters:

- For **Block Size**, type 82. For **Unknown**, open the list and select **US Survey Feet**.
- For **Block Point Limit**, keep the default of **8192**.

The Block Size and Block Point Limit parameters control how many points are in one block. To determine this value, it is important to consider the average point spacing, the objects of interest, available dedicated GPU memories, and the batch size when setting these two parameters. A general rule is that the block size should be large enough to capture the objects of interest with the least amount of subsampling.

You'll start with the Block Size of 82 feet (about 25 meters) and the Block Point Limit of 8,192. The measure 82 feet is a proper block size to capture the power line's geometry. You'll examine the output to see whether 8,192 is an appropriate block point limit. If most blocks have more than 8,192 points, you'll need to increase the block point limit to reduce subsampling.

Note: For the large data, it is recommended that you use the **LAS Point Statistics As Raster** tool to generate the histogram to determine the proper block size and block point limit before running the **Prepare Point Cloud Training Data** tool. To use the tool, choose **Point Count** as the method and **Cell Size** as the sampling type and test with different values for **Block Size**.

7. Click **Run**.

A confirmation message appears at the bottom of the tool pane when the tool is finished running.

8. Click **View Details**.

On the Messages tab, you can see two histograms of the block point count, one for the training data and one for the validation data.

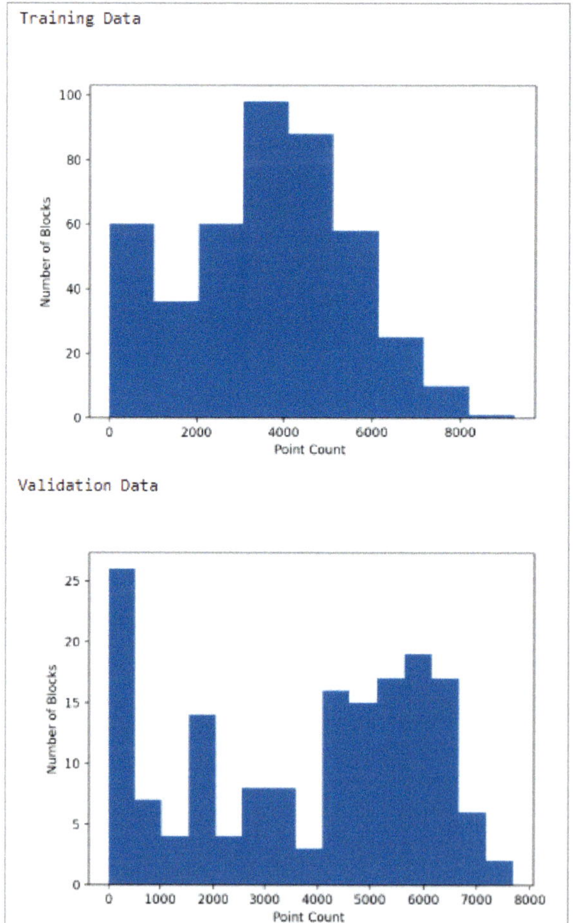

Both histograms show that almost all the blocks have a point count of less than 8,000, which confirms that a Block Size of 82 feet and the default Block Point Limit of 8,192 are proper for these datasets.

9 Close the details window.

10 Open File Explorer and browse to **GeoAI_Data\Chapter06\results**.

The results folder now contains a training_data_small.pctd folder.

11 Double-click **training_data_small.pctd** to open the folder.

The output file contains two subfolders, train and val, which contain the exported training and validation data, respectively.

12 Open each folder to review their contents.

In each folder, there is a Statistics.json file, a ListTable.h5, a BlockPointCount-Histogram.png file, and a 0 folder containing Data_x.h5 files.

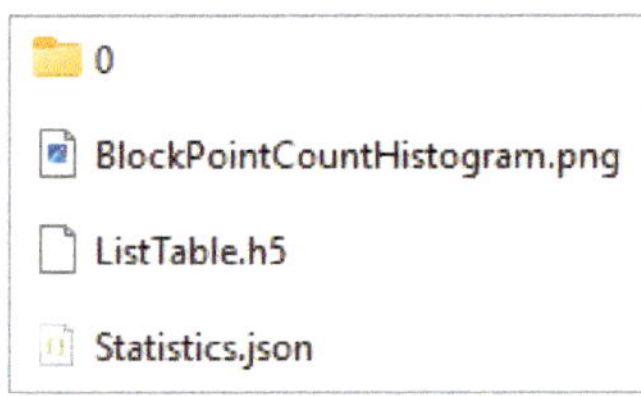

The ListTable.h5 and Data_x.h5 files include the information of points (xyz and attributes such as return number, intensity, and so on), which are organized into blocks.

Now that you have prepared the training data, you can use the blocks to train a classification model with deep learning.

Train a classification model

Next, you'll use the Train Point Cloud Classification Model geoprocessing tool to train a model for power line classification using the small training dataset. The results may vary based on different runs of the tool and different computer configurations.

1. In the **Geoprocessing** pane, search for and open the **Train Point Cloud Classification Model** tool and apply the following settings:

   - For **Input Training Data**, browse to **Chapter06\results** and double-click **training_data_small.pctd**.
   - For **Pre-trained Model**, leave it blank.
   - For **Model Architecture**, confirm **Point Transformer V3** is selected.

   > **Note:** You'll use **RandLA-Net** architecture to train the classification model. **RandLA-Net** uses random sampling to reduce memory usage and computational cost and uses local feature aggregation to preserve useful features from a wide neighborhood.

   - For **Attribute Selection**, check the box for **Intensity and Relative Height**.
   - For **Output Model Location**, browse to the **results** folder. Select the folder and click **OK**.
   - For **Output Model Name**, type Powerline_classification_model_small_data.
   - For **Minimum Points Per Block**, type 1000.

   The intensity of power lines is lower when compared with other features, so it is an effective attribute for distinguishing power lines. The relative height of power lines is usually within a certain range, so it is also used to separate power lines from other features.

   By setting the Minimum Points Per Block to 1,000 during the training, blocks whose points are less than 1,000 will be skipped. Blocks with a small number of points are likely to be located at the boundaries where there are no power line points. Skipping these blocks in the training will make the training faster and make the model's learning more efficient. This parameter is applied only on the training data blocks, not the validation data blocks.

2 Expand **Manage Classes** and apply the following settings:

- Under **Class Remapping**, for **Current Class**, select **14**. For **Remapped Class**, select **14**.
- In the next row, select **OTHER**, and from **Remapped Class**, select **1**.

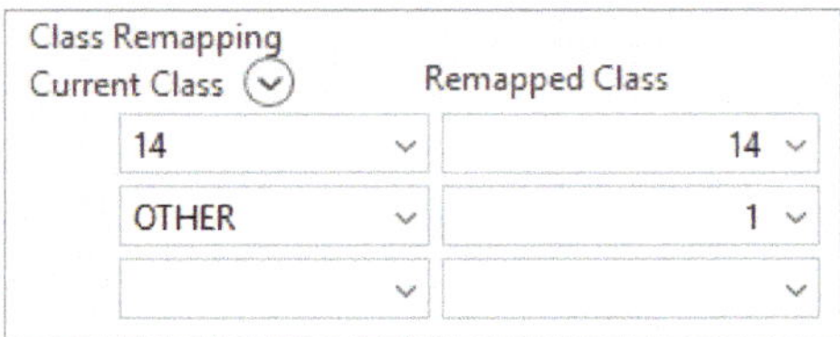

By specifying the Class Remapping, classification code 14, the class that represents wire conductors, will remain unchanged, meaning points in the LAS dataset that already classify as power lines will remain power lines. All other class codes (1, 5, and 15) will be remapped to the class code of 1. The resulting training data will have only two classes, 1 and 14, making the power lines easier to distinguish in the scene.

- Under **Class Description**, accept the populated class codes and descriptions.

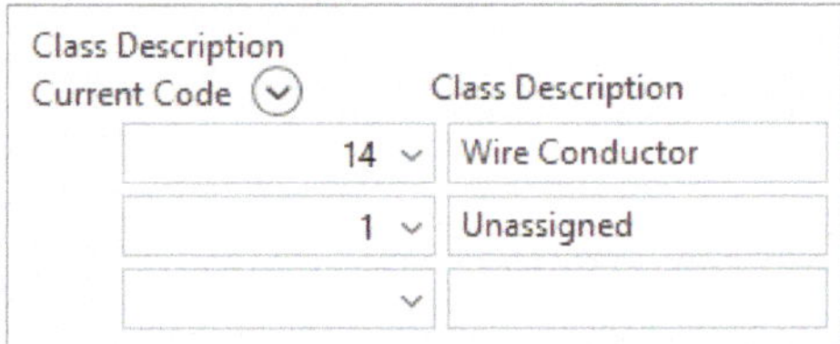

3 Expand **Training Parameters** and apply the following settings:

- For **Model Selection Criteria**, accept the default of **Recall**.

Model Selection Criteria specifies the statistical basis that will be used to determine the final model. The default of Recall will select the model that achieves the best macro average of the recall for all class codes. Each class code's recall value is determined by the ratio of correctly classified points (true positives) over all the points that should have been classified with this value (expected positives). For each class, its recall value is the ratio of correctly predicted points of this class to all the reference points of this class in the validation data. For example, the recall value of class code 14 is the ratio of correctly predicted power line points to all the reference power line points in the validation data.

- For **Maximum Number of Epochs**, type 10.

  An epoch is the complete cycle of the entire training data learned by the neural network (in other words, the entire training data is passed forward and backward through the neural network one time). You'll train the model for 10 epochs to save time.

- For **Iteration Per Epoch (%)**, accept the default of **100**.

  Leaving the Iteration Per Epoch (%) parameter at 100 ensures that all the training data will be passed per epoch.

> **Note:** You can also pass a percentage of the training data at each epoch. Set this parameter to something other than 100 if you want to reduce the completion time for each epoch by randomly selecting fewer batches. However, this can lead to more epochs before the model converges. This parameter is useful if you want to quickly see the model metrics in the tool's message window.

- For **Learning Rate Strategy**, keep the default of **One Cycle Learning Rate**.

- For **Learning Rate**, type .005.

> **Note:** You can also leave **Learning Rate** blank and allow the tool to find an optimal learning rate for you.

- For **Batch Size**, type 6.

  Batch Size specifies how many blocks are processed at a time. The training data is split into batches. For example, if the Batch Size value is set to 20, 1,000 blocks are split into 50 batches, and each of the 50 batches are processed in one epoch. By splitting the blocks into batches, the process will consume less GPU memory.

> **Note:** If you received a CUDA out of memory error, use a smaller batch size.

- Uncheck the box for **Stop training when model no longer improves**.

  Unchecking this option will allow the training to run for 10 epochs. If checked, the training will stop when the model is no longer improving after several epochs, regardless of the maximum number of epochs specified.

4. Click **Run**.
5. At the bottom of the **Geoprocessing** pane, click **View Details**. Then, click the **Messages** tab.

Note: The geoprocessing messages populate as the tool runs. When the tool is finished running, the geoprocessing message shows the results for each epoch.

378 of 437 training data blocks will be processed.
166 validation data blocks will be processed.

Epoch	Training Loss	Validation Loss	Accuracy	Precision	Recall
0	0.0618295	0.0783559	0.99673	0.981509	0.965653
1	0.0163609	0.0281205	0.99511	0.92283	0.948014
2	0.0177829	0.0139706	0.99655	0.954223	0.952841
3	0.00184092	0.011109	0.99694	0.974221	0.936905
4	0.00383078	0.0116663	0.99727	0.96628	0.950527
5	0.00545739	0.0159347	0.99702	0.96979	0.941741
6	0.000364895	0.0155672	0.99722	0.975551	0.939779
7	0.0019259	0.00942634	0.9977	0.973626	0.950997
8	0.000624984	0.0061022	0.99835	0.99198	0.975173
9	0.00133948	0.00947594	0.99762	0.97165	0.952875

Note: The information in your messages may be different from the example shown.

The tool reports the following information:

- GPU used in the training.
- Count of the training data blocks used in the training (only training data blocks containing more than 1,000 points are used in the training).
- Count of the validation data blocks used in the validation—all the validation data blocks are used in the validation.
- Iteration per epoch—the count of the training data blocks divided by the batch size.

The tool reports Training Loss, Validation Loss, Accuracy, Precision, Recall, F1-Score, and Time spent for each epoch. As each epoch progresses, you can see the Training Loss and Validation Loss values decrease, which indicates that the model is learning.

This model is not an accurate model, which is expected because the model was trained with a small dataset. These results highlight the need to have more sample points in separate datasets so that you can achieve better results.

6. Close the details window.

Examine the training outputs

Next, you'll look at the results of training the classification model.

1. In **File Explorer**, browse to **C:\GeoAI\Chapter06\results**.

   The training and validation datasets each have two subfolders. One is the model folder, and another is the checkpoints folder.

2. Open the **Powerline_classification_model_small_data** folder.

   The Powerline_classification_model_small_data folder includes the saved model at the epoch of 9. The model folder contains several files. Among them, the .pth file is the model file, and the .emd file is the Esri Model Definition file (a configuration file in JSON format). The model_metrics.htm file includes a learning loss graph. The .dlpk file is the deep learning model package. It is a compressed file and can be shared in ArcGIS Online.

3. Open the **ModelCharacteristics** folder.

   The ModelCharacteristics folder contains a loss graph and ground truth and predictions results graph.

4. Go back to the **results** folder. Open the **Powerline_classification_model_small_data.checkpoints** folder and view its contents.

   The Powerline_classification_model_small_data.checkpoints folder includes the models folder, which contains a model for each of the epoch checkpoints and two .csv files. When the training tool was running, one checkpoint was created after each epoch. Each checkpoint contains a .pth file and an .emd file.

5 In the **Powerline_classification_model_small_data.checkpoints** folder, open the **Powerline_classification_model_small_data_Epoch_Statistics** Microsoft Excel file to see the statistics.

A	B	C	D	E	F	G	H	I
EPOCH	TRAINING	VALIDATI(	ACCURAC'	PRECISIOI	RECALL	F1_SCORE	ELAPSED_TIME	
0	0.302041	0.276751	0.99266	0.892824	0.966769	0.91703	0:01:46	
1	0.047826	0.04812	0.99725	0.952568	0.968227	0.9599	0:01:45	
2	0.039525	0.019089	0.99683	0.950448	0.957163	0.952664	0:01:46	
3	0.009068	0.012169	0.99757	0.949437	0.976464	0.961038	0:01:44	
4	0.008416	0.01066	0.9971	0.97311	0.942039	0.948482	0:01:39	
5	0.007679	0.009562	0.99719	0.963009	0.943423	0.944873	0:01:32	
6	0.013038	0.006951	0.99789	0.964422	0.960265	0.961664	0:01:33	
7	0.002113	0.007539	0.99729	0.965808	0.951671	0.951013	0:01:31	
8	0.00136	0.005449	0.99867	0.990803	0.9824	0.985366	0:01:31	
9	0.005013	0.005428	0.99848	0.963425	0.975821	0.968522	0:01:31	

The CSV file opens in Excel.

Note: If the Microsoft Excel window appears prompting you to convert large numbers into scientific notation, click **Convert.**

It is the same table that you saw in the tool's message.

The Powerline_classification_model_small_data_Statistics file includes the precision, recall, and f1-score values for each class after each epoch. In certain cases, the model with the best overall metrics may not be the model that performed the best in classifying a specific class code. If you are interested in classifying only certain class codes, you may consider using the checkpoint model associated with the best metrics for that class code. Viewing the statistics in Excel allows you to sort the columns and easily locate the epoch with the highest recall value.

You have trained a model using a smaller sampling of points. Next, you'll use a model that was trained using a larger sampling of points to classify a LAS dataset.

Tutorial 6-2

Explore model results and choose the best epoch

You will classify the LAS dataset containing more than 3 million points using a trained model. Classifying the LAS dataset using a trained model allows you to locate power lines in the study area for risk assessment analysis. When the model was trained, two classification codes were designated, Unassigned and Wire Conductor. By classifying points in the point cloud to wire conductor and the rest as unassigned, it will make the dataset more valuable, as it will clearly identify power lines.

You will explore the results of a provided model training that used the large dataset. You will look for the best recall value for wire conductors.

1. Open File Explorer and browse to **C:\GeoAI\Chapter06\results**.
2. Open the **Powerline_classification_model_large_data.checkpoints** folder and double-click **Powerline_classification_model_large_data_Statistics** to view the statistics in Excel.

   The file opens in Excel.
3. In Excel, double-click the column dividers to expand them so that you can see all the text.

	A	B	C	D	E	F
1	EPOCH	CLASS_CODE	CLASS_DESCRIPTION	PRECISION	RECALL	F1_SCORE
2	0	1	Unassigned	0.908717554	0.999622203	0.951989322
3	0	14	Wire Conductor	0.990488739	0.288281157	0.445502016
4	1	1	Unassigned	0.998148663	0.998441761	0.998294264
5	1	14	Wire Conductor	0.958082944	0.950081375	0.953435094
6	2	1	Unassigned	0.999238431	0.998703438	0.998970345
7	2	14	Wire Conductor	0.964261536	0.979275692	0.971275565
8	3	1	Unassigned	0.999255505	0.998313948	0.998783802
9	3	14	Wire Conductor	0.954034876	0.979420959	0.966053087

   You will sort the Recall column and find the largest value for Wire Conductor and use that epoch for classifying the LAS dataset. The largest Recall value for the Wire Conductor indicates the epoch that you should use in the classification.
4. Click column header **E** to highlight the entire **RECALL** column.

5. On the ribbon, click the **Home** tab. In the **Editing** group, click **Sort & Filter** and select **Sort Largest to Smallest**.

6. In the **Sort Warning** window that appears, click **Sort**.

   The rows in the spreadsheet are sorted so you can see the highest value for Wire Conductor.

7. Locate the first row with a **CLASS_CODE** of **14** and a **CLASS_DESCRIPTION** of **Wire Conductor** (row 23).

   This is the highest recall value of 0.993655113, for Wire Conductor, which occurred in epoch 18.

21	1	1	Unassigned	0.998148663	0.998441761	0.998294264
22	3	1	Unassigned	0.999255505	0.998313948	0.998783802
23	18	14	Wire Conductor	0.987222604	0.993655113	0.990359409
24	19	14	Wire Conductor	0.99307506	0.992359409	0.992703749
25	13	14	Wire Conductor	0.99351991	0.991258864	0.992378093

   You will use epoch, or checkpoint, 18, when you classify the LAS dataset because it has the highest recall value for the power line class code and will provide the best results.

Classify power lines using the trained model

Now you will use a trained model to classify the power lines from a test dataset. You will apply processing boundaries to classify points only within the boundaries.

1. Return to ArcGIS Pro. In the **Catalog** pane, expand **Folders** > **Chapter06** > **testdata**. Right-click **test.lasd**, click **Properties**, and update the statistics. Click **OK**.

2. Right-click **test.lasd** and click **Add To New** > **Local Scene**.

> **Note:** LAS datasets can be large and take more time to display than other layers. To speed up the display, you can build pyramids using the **Build LAS Dataset Pyramid** tool. This tool creates or updates a LAS dataset display cache, which optimizes its rendering performance.
>
> If you have white spaces in your LAS dataset layer, this could be a caching issue. You can go to the properties of the LAS dataset, select **Cache**, and click **Clear Cache** to fix this, or it will be fixed automatically when you close and open ArcGIS Pro.

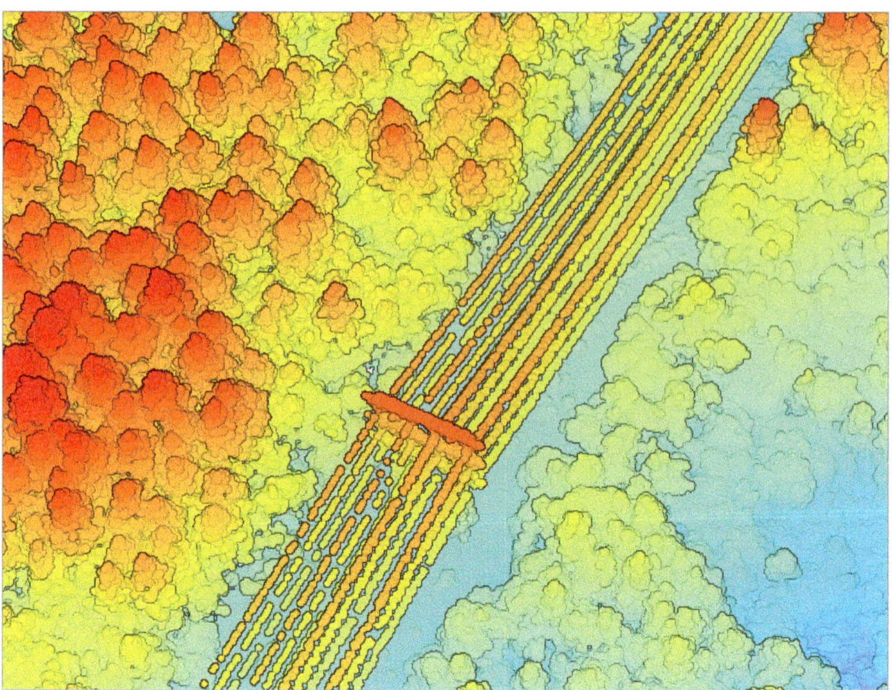

The test.lasd LAS dataset appears in a local scene.

3 In the **Contents** pane, click **test.lasd** to select it. On the ribbon, click the **LAS Dataset Layer** tab, click the drop-down arrow for **Symbology**, and select **Class**.

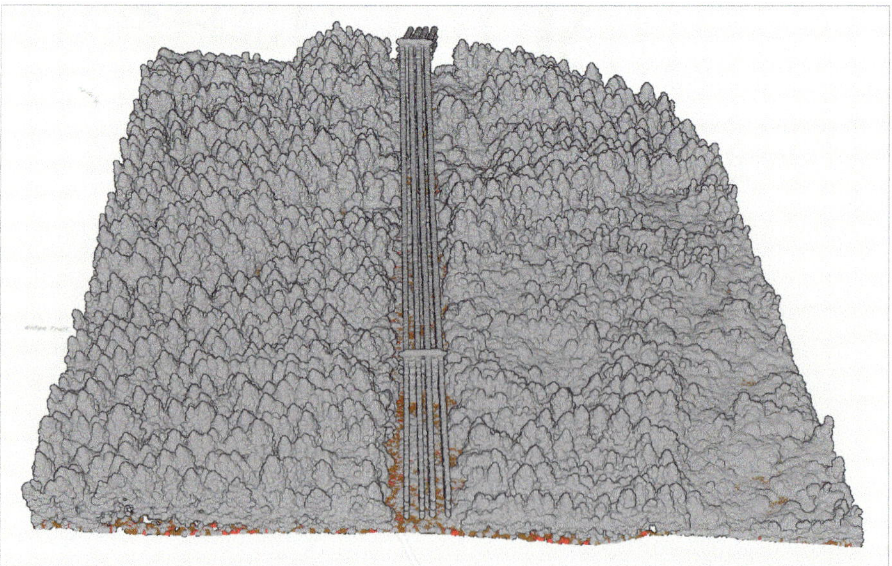

The test.lasd LAS dataset is now rendered using class codes. Ground points (class 2) and low noise points (class 7) are already classified. The remaining points are unassigned (class 1).

4 In the **Geoprocessing** pane, search for and open the **Classify Point Cloud Using Trained Model** tool. Apply the following settings:

- For **Target Point Cloud**, select **test.lasd**.
- For **Input Model Definition**, browse to **Chapter06\results\Powerline_classification_model_large_data.checkpoints\models** and open the **checkpoint_2023-03-28_18-21-34_epoch_18** folder. Select **checkpoint_2023-03-28_18-21-34_epoch_18.emd** and click **OK**.

> **Note:** You can choose an .emd file, a .dlpk file, or the URL to the model shared at ArcGIS Online or ArcGIS Living Atlas of the World.

- Leave the **Batch Size** blank and allow the tool to calculate an optimal batch size for you based on the available GPU memory.
- For **Processing Boundary**, browse to **Chapter06\testdata\boundaries.gdb** and double-click **bnd_test**.
- For **Reference Surface**, browse to **Chapter06\testdata**. Click **dem_test.tif** and click **OK**.

> **Note:** Because the power line classification model was trained using boundaries, a reference surface, and excluding certain class codes, the same parameters should also be set for the **Classify Point Cloud Using Trained Model** tool.

- For **Excluded Class Codes**, type 2. Click **Add another** and type 7.

  Since class codes 2 and 7 are excluded from the classification, the model classifies only the remaining points (class 1). When the model runs, the points are predicted as power lines and will have their class codes assigned as 14; otherwise, their class codes are kept as 1.

5 Click **Run**.

The tool runs. Next, you will update the symbology to show the results of the classification model.

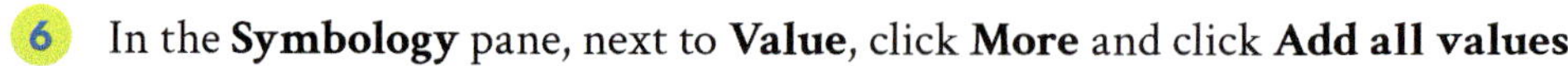

6 In the **Symbology** pane, next to **Value**, click **More** and click **Add all values**.

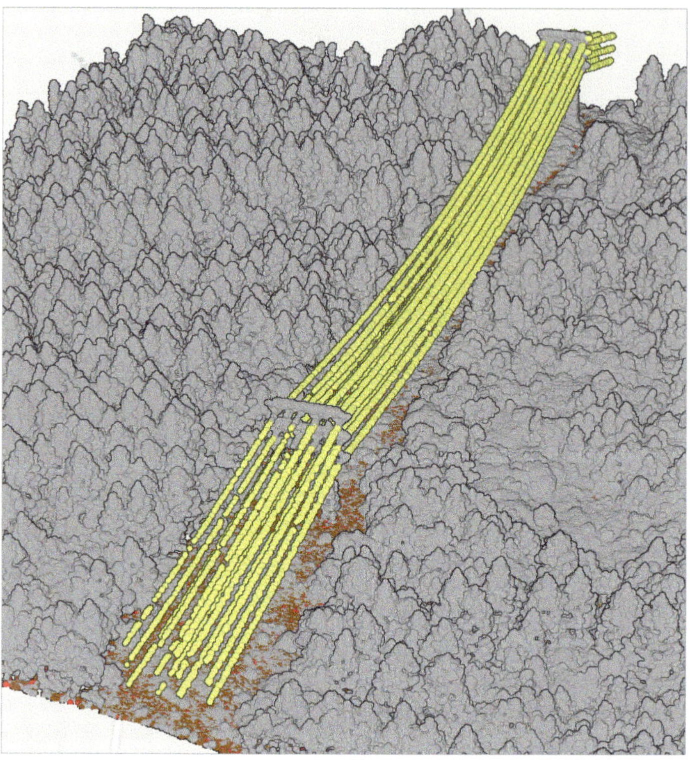

The symbology updates and the points in the point cloud that are wire conductors display in yellow.

7 Use the navigation tools and shortcuts to examine the classification results in the local scene.

The classification results are accurate using the trained model. Almost all the power line points are correctly classified. Notice that the power poles are still Unassigned.

Take the next step (optional)

You can use the **Extract Power Lines From Point Cloud** tool to generate 3D lines to model the power lines. You could also use the **Generate Clearance Surface** tool to determine the clearance zone around the power lines. The points within the clearance zone, mostly trees, could cause a power outage or spark a fire. These tools can provide important information for insurance risk assessments.

Summary

In this tutorial, you learned the workflow of point cloud classification using deep learning technology. You gained experience with deep learning concepts, the importance of validation data in the training process, and how to evaluate the quality of the trained models.

This chapter is based on an Esri tutorial by Jie Chang and Lindsay Weitz.

CHAPTER 7

Training a model using automated deep learning

Objectives

- Train multiple deep learning models.
- Analyze the model training report to determine the best-performing model.
- Use deep learning to classify land cover in SAR imagery.

Introduction

Changes in land cover over time can be tracked using optical remote sensing techniques. In this chapter, you'll prepare and train a deep learning model to classify land use and land cover using Sentinel-1 imagery from 2018. You'll use the Train Using AutoDL tool to train several deep learning models and identify the best-performing one. Because you are working with land-cover classification, you'll use the trained model on 2024 SAR imagery of the same area.

Requirements

- ArcGIS Pro
- ArcGIS Image Analyst
- Deep Learning Libraries for ArcGIS Pro
- Recommended: NVIDIA GPU with a minimum of 8 GB of dedicated memory

Note: For this tutorial, an NVIDIA GPU with a minimum of 8 GB of dedicated memory is recommended. Based on whether your computer has a GPU and what its specifications are, this process may take from under 2 minutes to 20 minutes or more.

If you are not sure whether your computer has a GPU and what its specifications are, see chapter 1.

Tutorial 7-1

Identify the best-performing model

Land Use Land Cover (LULC) classification provides an overview of the general categories of land use and land cover for broad geographic areas, based on remotely sensed imagery. It plays a vital role in various aspects, such as environmental monitoring, resource management, biodiversity conservation, disaster risk reduction, and climate change analysis. It facilitates the systematic monitoring of changes in land use, efficient allocation of resources, preservation of habitats, identification of areas prone to disasters, and evaluation of climate change impacts.

Set up the project

1. In File Explorer, browse to **C:\GeoAI_Data** and open the **Chapter07** folder. Double-click **AutoDL_tutorial.aprx** to open the project in ArcGIS Pro.

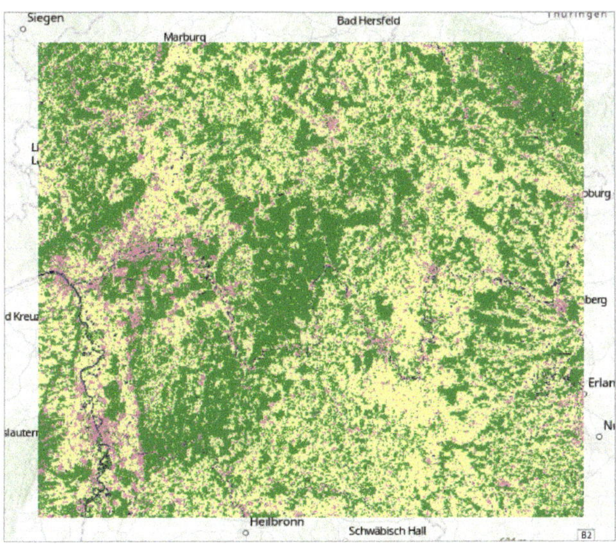

A map appears, showing part of Germany. An image layer, LULCRaster2018.tif, displays on top of the topographic basemap.

The LULCRaster.tif layer is a classified raster showing Level 1 LULC classes. This classification system provides a broad categorization of the earth's surface into general land-use and land-cover types, such as urban areas, agricultural land, forests, water bodies, and wetlands. It serves as the most basic level of classification, providing an overview of major land categories for large-scale analysis.

2 On the ribbon, on the **Map** tab, in the **Navigate** group, click **Bookmarks** and click the **Speyer** bookmark.

The map zooms to the southwestern part of the LULCRaster2018.tif layer. The layer shows a built-up area near a river, with forest and agricultural areas.

3 In the **Contents** pane, uncheck the box for **LULCRaster2018.tif** to turn off layer visibility. Check the box for **SARImagery2018.tif** to turn on its visibility.

The SARImagery2018.tif layer is derived from remotely sensed Sentinel-1 Ground Range Detected (GRD) SAR imagery from 2018. This layer has a 10 m resolution and is stored in TIFF format with three bands.

> **Note:** The original SAR imagery was downloaded and processed to prepare it for analysis. Originally consisting of two polarization bands, VV and VH, it was downloaded, and using the **Band Arithmetic** raster function, the VV/VH derived band was created. This band consists of the VV band divided by the VH band. This band combination is useful because it highlights differences in scattering behavior, which allows you to make inferences about the surface characteristics. After this processing, the **Composite Bands** geoprocessing tool was used to create a composite of VV, VH, and VV/VH with 8-bit unsigned pixel depth.

Sentinel-1 data is collected regularly, which allows comparison of different images to detect change over time.

Manually classifying all pixels of this imagery into Level 1 LULC classes would be a long and tedious process. A deep learning model can automate LULC classification of imagery, which can help you regularly classify new data and detect changes. In this chapter, you will determine whether you can use this workflow to annually update your land-use land-cover data and report changes over time.

Explore the training data

The project contains the training samples prepared using the Export Training Data For Deep Learning tool. You'll use the samples to train and identify the best-performing model.

1. In the **Catalog** pane, expand **Folders** > **Chapter07** > **trainingdata**.

   The training data folder contains the training data that you will use.

2. Explore the contents of the **images** and **labels** folders.

   - images—contains image chips, extracted from the SARImagery2018.tif layer using the Export Training Data For Deep Learning tool.
   - labels—contains label images showing the classified land-cover types.

   Training a deep learning model to classify this imagery involves presenting the model with the image chips and with the matching labels, allowing the model to learn which classes are associated with which SAR band combinations.

Train multiple models

One of the challenges with deep learning is determining which model architecture to use. This process of choosing and training a model can be confusing and lengthy, as the models have different strengths and weaknesses, and they require different inputs and parameters. The Train Using AutoDL tool allows you to select a set of model architectures to train. It then trains and tests them and reports which model performed best.

The Train Using AutoDL tool trains the set of deep learning models by building training pipelines and automating much of the training process. This includes data augmentation, model selection, hyperparameter tuning, and batch size deduction. Its outputs include performance metrics of the best model on the training data and a trained deep learning model package (.dlpk file) that can be used in the Extract Features Using AI Models or Classify Pixels Using Deep Learning tools to classify other images.

1. On the ribbon, click the **Analysis** tab. In the **Geoprocessing** group, click **Tools**.

2. In the **Geoprocessing** pane, search for and open the **Train Using AutoDL (GeoAI Tools)** geoprocessing tool and apply the following settings:

   - For **Input Training Data**, click the **browse** button. Browse to **Folders** > **Chapter07**, click the **trainingdata** folder, and click **OK**.
   - For **Output Model**, browse to **Folders** > **Chapter07**, type LULCClassifierModel, and click **Save**.

     This creates a new folder to contain the output trained model or models.

   - For **Total Time Limit (Hours)**, type 4.

     The tool will work on the task for up to four hours. Depending on your computer's GPU, it may use the entire time, or it may complete the task in a shorter time.

     Based on the format of the training data, you'll see a list of supported neural networks specific to pixel classification.

3. Expand the **Advanced Options** section and apply the following settings:

   - For **Neural Networks**, click the **Add Many** button. In the list that appears, check the boxes next to **HRNet**, **PSPNetClassifier**, and **UnetClassifier**. Click **Add**.

     These are neural networks that classify pixels in a raster using semantic segmentation. They are commonly used for land-cover classification.

   - Check the box for **Save Evaluated Models**.

     You have specified that the Train Using AutoDL tool should run for four hours training and evaluating the three models.

4. In the **Train Using AutoDL** tool, click the **Environments** tab. For **GPU ID**, type 0.

   If your CUDA-enabled GPU has a different GPU ID, use that ID number. This may be necessary when your computer has more than one GPU.

5. Click **Run**.

   The process will run for up to four hours. You can view messages about the status of the process as the tool runs.

Note: If you have a computer with a suitable GPU, you can run the tool to train and evaluate the three models for up to four hours.

Optionally, you can skip the training step and review a folder that has been prepared for you with all the outputs of the tool. If you are not going to run the model training process, read the next three steps and start working again in the next section, "Review the Model Training Results."

6. At the bottom of the **Train Using AutoDL** pane, click **View Details**.

7. Click the **Messages** tab.

When the process is complete, you can view the outputs in the Messages pane.

```
GPU is being used.
Given time to process the dataset is: 4.00 hours.
Number of images that can be processed in the given time: 377.
Time required to process the entire dataset of 377 images is 1.53 hrs.
Starting the training process.
```

Model	Train Loss	Valid Loss	Accuracy	Dice	Learning Rate	Time	Backbone
PSPNetClassifier	0.744	0.684	0.83144	0.831	0.0014454	0:11:49	resnet50
UnetClassifier	0.699	0.642	0.75981	0.760	0.0000525	0:06:03	resnet34
HRNet	1.067	0.986	0.66279	0.663	0.0020893	0:01:18	resnet34

In the training process, the tool randomly selects 10 percent of the training dataset to reserve for validation and trains the models on the other 90 percent. As training proceeds, the model calculates how well the predicted values match the values in the validation dataset. The table summarizes how well each model performed. Partly because of this random selection of validation samples, training models with this tool is not deterministic. The tool also randomly sets some initial conditions each time it is run.

Given the same set of training data, different models may be chosen as the best model, and different values will appear in this table.

The table includes columns for both Training Loss and Validation Loss. Training Loss shows how well the model learned. Validation Loss shows how well what the model learned performed on the validation set, in effect, showing how generalized the learning was. Lower values for these two measures indicate better training performance, so in this case, the UnetClassifier model performed best in terms of Training Loss and Validation Loss.

The PSPNetClassifier model had a higher Accuracy value than the other models. For accuracy, higher values are better.

Accuracy and dice are measures of how well the model correctly classified pixels. The PSPNetClassifier model also had a higher Accuracy value than the other models.

The Learning Rate is a hyperparameter used in the training of the neural networks. If you did not specify the value, it will be calculated by the training tool. The tool attempts to optimize the learning rate, weighing speed against quality. The resulting Learning Rate value listed in the table is primarily of interest if, as an advanced user, you want to continue training the model and need guidance on choosing the Learning Rate value for that additional training.

The Time column indicates the time it takes to train each model. You will notice that the time is larger for the first model than for the others. This is because there is some data processing overhead that is done for the first model and is then reused by the models that follow.

The Backbone column is the default model backbone. If you set the AutoDL Mode parameter to Advanced instead of Basic, multiple backbones may be tried.

8. Close the **Messages** pane.

Review the model training results

The project package you downloaded includes a zipped folder of the Train Using AutoDL tool results. These results may be different from the ones you got from running the tool on your own.

1. In the **Catalog** pane, browse to **Folders** > **Chapter07**. Right-click the **userdata** folder and click **Copy Path**.
2. In File Explorer, paste the path into the path box.

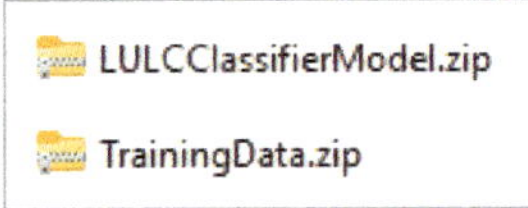

Inside this folder are two zip archives. The LULCClassifierModel.zip archive contains the results created by running the Train Using AutoDL tool with the settings specified earlier.

The TrainingData.zip archive contains the data used to create the training data.

3 Right-click the **LULCClassifierModel.zip** archive and click **Extract All**. Then, click **Extract**.

4 Open the **LULCClassifierModel** folder.

This folder contains several outputs from running the tool. These include the following:

- ModelCharacteristics—a folder containing images used in the README.html file.
- ArcGISImageClassifier—a Python script with code used in classifying imagery for the training process.
- LULCClassifierModel.dlpk—a complete package of all the files stored in the model output folder including the trained model, the model definition file, and the model metrics file. This package can be shared to ArcGIS Online and ArcGIS Enterprise as a trained model item for others to use.
- LULCClassifierModel.emd—a model definition file that contains model information about the tile size, classes, model type, and so on.
- LULCClassifierMode.pth—a pretrained weights file, usually saved in a PyTorch format.
- model_metrics.html—an HTML page with details about the learning rate used and the accuracy of the trained model.
- README.html—an HTML page with details about the evaluation of models and accuracy of the best-performing model.

5 Double-click the **README.html** file.

The page opens on a browser tab. This page shows information about the best-performing model, how it compared with the other models, and how well it was able to classify LULC from your input training data.

6 Scroll down to the **Training and Validation loss** report.

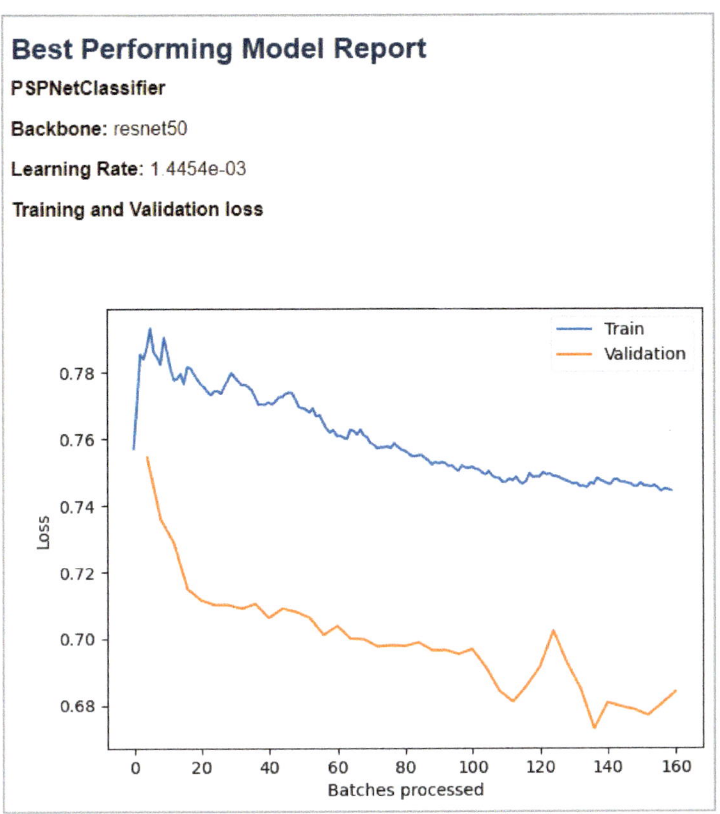

The Training and Validation loss section displays a graph of the amount of error that was present as the model trained over time. When the tool ran, 90 percent of your input data was used to train the model and 10 percent was used to validate the model to determine its accuracy. Ideally, you would see these loss values decrease and converge as the number of images (batches) processed increases over the course of available time.

> **Note:** Because the validation samples are randomly selected from the set of training image chips and because some hyperparameters are randomly set to start training, the **Training Loss** and **Validation Loss** metrics can be different each time the tool is run, even on the same training dataset.

In this graph, you can see that for the first 60 batches of images processed, the error shown by the Validation line is high but varies quite a bit for each batch. After 60 batches, the amount of error in the Validation process decreases and varies less, except for a peak at 120. The Train line shows a more steadily decreasing amount of error.

7 Scroll down to the **Analysis of the model** table.

Analysis of the model

Per class metrics:

	NoData	Artificial surfaces	Forest and natural areas	Agricultural areas	Water bodies
precision	0	0.711454	0.860713	0.876042	0.773818
recall	0	0.825857	0.830766	0.834371	0.838877
f1	0	0.764399	0.845474	0.854699	0.805035

The Analysis of the model section displays the precision of the classes of data. Your model technically had five classes: four for land cover and one for NoData. A higher precision value means the model is more confident in its results.

Note: You can read more about interpreting the precision and accuracy statistics of deep learning tools at link.esri.com/ExploreGeoAI/ObjDetect.

This page also shows a few sample chips comparing your original LULC training data, Ground Truth, on the left and the model's Predictions on the right. Ideally, the prediction should closely match the original ground truth.

8 Close the **README** page on your web browser.

You've trained a deep learning model for LULC Classification on Sentinel-1 imagery taken in 2018 and found that the best-performing model for that task is based on the UnetClassifier architecture. Next, you'll use this model to automatically classify land cover in Sentinel-1 imagery taken in 2024.

Tutorial 7-2

Apply the model

Once a deep learning model is created, it can be used to quickly classify land cover on similar data captured on different dates. This allows you to monitor land-cover change over time. In this tutorial, you'll take the model that you created and use it to classify Sentinel-1 imagery captured in 2024 in the same geographic area.

Use the trained model to classify new imagery

1. In ArcGIS Pro, click the **Deploy Model** map tab.

   The Deploy Model map shows the SARImagery2024.tif layer.

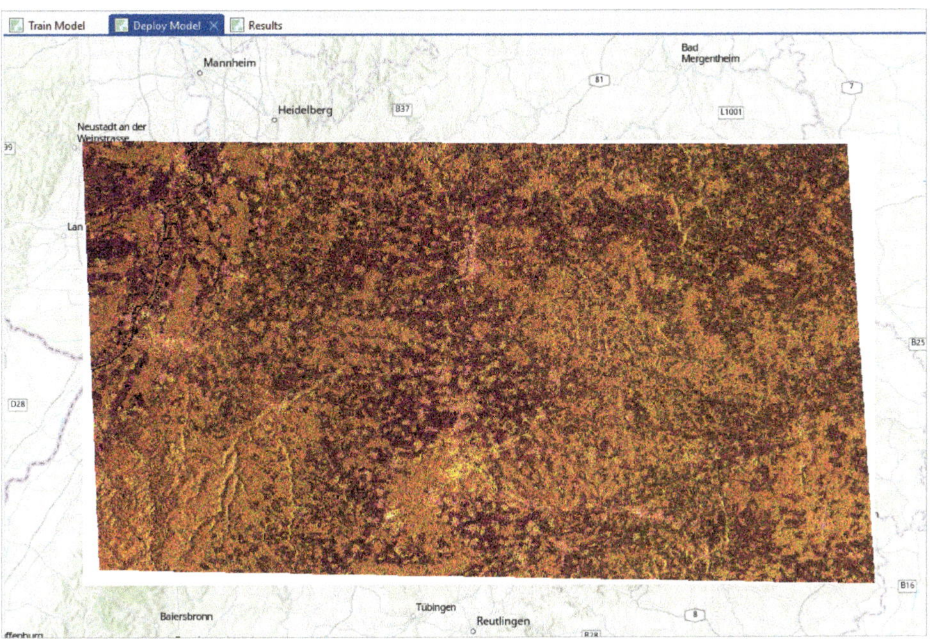

2. In the **Geoprocessing** pane, search for and open the **Classify Pixels Using Deep Learning** tool and apply the following settings:

   - For **Input Raster**, select **SARImagery2024.tif**.
   - For **Output Raster Dataset**, type ClassifiedLULC2024.
   - For **Model Definition**, click the browse button and browse to the **Chapter07\ LULCClassifierModel** folder. Click the **LULCClassifierModel.dlpk** deep learning package.

> **Note:** If you did not train the model on your machine, you can use the trained model that is provided with the project. Browse to the **Chapter07\userdata \LULCClassifierModel\LULCClassifierModel** folder and click the **LULCClassifierModel.dlpk** deep learning package.

After the deep learning package is loaded by the tool, the model Arguments table appears. You'll accept the default values.

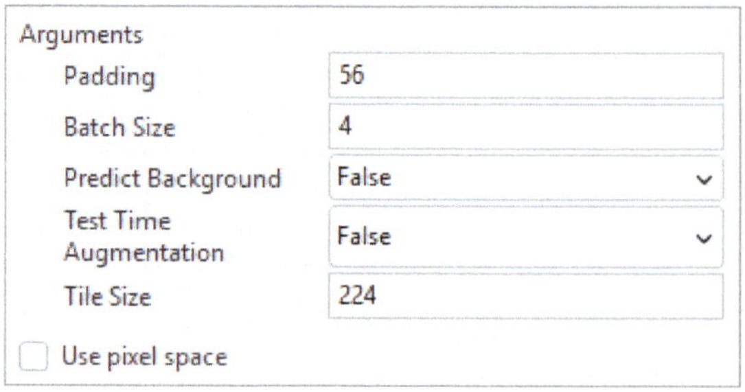

Arguments

Padding	56
Batch Size	4
Predict Background	False
Test Time Augmentation	False
Tile Size	224

☐ Use pixel space

Note: The ability to process larger batch sizes is based on your computer's GPU. If you have a powerful GPU, you can process up to 64 training samples at a time. If you have a less powerful GPU, you may have to set the **Batch Size** parameter to 8, 4, or 2 and rerun the tool.

3. Click the **Environments** tab. Apply the following settings:

   - In the **Raster Analysis** section, for **Cell Size**, type 10.

     The cell size of the SAR data doesn't exactly match the cell size that the model was trained on, so you can specify that the output should have a cell size of 10.

   - For **Processor Type**, choose **GPU**.

     The process of classifying this image may take 40 minutes or more. Optionally, you can skip running the tool and view the tool output, which is provided in the project package.

   - In the **Processor Type** section, for **GPU ID**, type 0.

     Note: If your CUDA-enabled GPU has a different GPU ID, use that ID number. This may be necessary when your computer has more than one GPU.

4. Optionally, click **Run**.

   If you run the tool, when it finishes, view the results on the map.

Note: The image class colors will be randomly assigned. You can change them to suit your preference. Right-click a class symbol, and in the color palette, choose a color of your liking.

5. If you didn't run the tool, click the **Results** map tab to see tool results.

   The classified imagery appears.

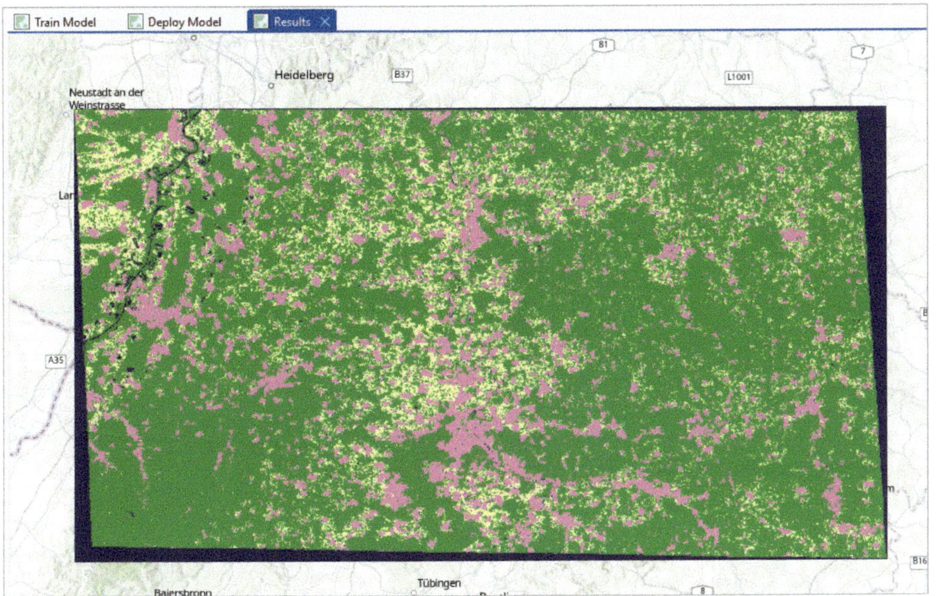

> **Note:** Deep learning is not a deterministic process, so the results you obtain may be slightly different.

You can compare the 2024 raster to the 2018 raster to detect large-scale land-use change over time. Now that the deep learning model has been trained, you can apply it to new SAR imagery every year, or more frequently. This trained model can become part of an effective workflow to monitor land-cover change over time.

Summary

In this tutorial, you used the Train Using AutoDL tool to train multiple models to classify land cover from Sentinel-1 SAR imagery and automatically identify which performed best. You then applied the best-performing trained model to more recent imagery.

This chapter is based on an Esri tutorial by Priyanka Tuteja.

CHAPTER 8

Classifying land cover with a pretrained deep learning model in ArcGIS Online

Objectives

- Create a classified raster from a tiled imagery layer.
- Use land-cover classification tools in ArcGIS Online.

Introduction

High-resolution land-cover layers are valuable tools for mapping and understanding the environment. They provide detailed information about the different types of land cover—such as vegetation, buildings, water bodies, and roads—at a fine-grained spatial resolution. One approach to creating such layers is applying GeoAI to drone imagery, classifying the imagery pixels into their corresponding land-cover types. Although it is possible to train your own deep learning model for this task, you can also take advantage of a pretrained model provided by ArcGIS Living Atlas of the World.

In this chapter, you'll use ArcGIS Online to run a pretrained model in the cloud and add the resulting layer to a web map.

Requirements

- ArcGIS organizational account with Professional or Professional Plus user type
- 1 credit

Tutorial 8-1

Review tiled imagery

Set up the project

You'll open ArcGIS Online Map Viewer and add the imagery data needed for the workflow.

1. In your browser, browse to www.arcgis.com. Sign in to your ArcGIS organizational account.
2. On the ribbon, click the **Map** tab.
3. On the **Contents** (dark) toolbar, click **Save and open** > **Save as**.
4. In the **Save map** window, for the **Title**, type Alexandra_Land_Cover. For **Folder**, save it to **ExploringGeoAI** (create as a new folder if you have not already). Click **Save**.

   You'll add drone imagery representing an area in Alexandra, South Africa. The imagery is high resolution, with each pixel representing a square of about 90 by 90 centimeters on the ground. It was captured by South Africa Flying Labs and resampled from 2.5 centimeters. The layer is stored in ArcGIS Online as an image tile service.

5. In the **Layers** pane, click the **Add** button. Click **My content**, and in the list, select **ArcGIS Online**.

6. In the search box, type Alexandra Orthomosaic 90 cm Tiled and press Enter. In the list of results, for the **Alexandra_OrthoMosaic_90cm** tiled imagery layer owned by **Esri_Tutorials**, click the **Add** button.

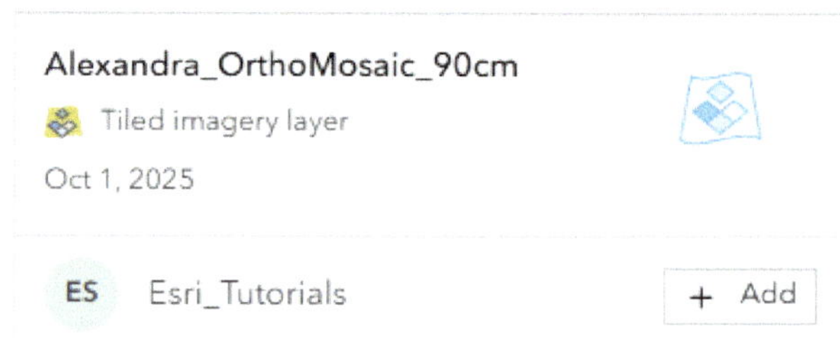

7. Zoom in and pan around the map to inspect the imagery more closely.

This True Ortho image layer was derived from multiple drone images.

8. Zoom out to ensure that the whole tile layer is visible on the map.

Tutorial 8-2

Run land-cover classification

Next, you'll open the tool to extract land cover.

Open the deep learning tool

You'll use the Classify Pixels Using Deep Learning tool for the analysis.

1. On the **Settings** (light) toolbar, click the **Analysis** button and click **Tools**.

   The Tools pane appears.

2. Search for and open the **Classify Pixels Using Deep Learning** tool.

   Note: When using analysis tools in ArcGIS Online, click **Estimate Credits** before running a tool to see the estimated number of credits it will use.

Set the tool parameters and choose a pretrained model

The Classify Pixels Using Deep Learning tool allows you to run a variety of deep learning models. You'll use a pretrained model from ArcGIS Living Atlas to classify this image. When using deep learning models to classify imagery, it is important that the model be trained on imagery that is a good match for the imagery you want to analyze. The number of bands, bit-depth, and resolution of the imagery used for training should match your imagery.

1. In the **Classify Pixels Using Deep Learning** tool pane, for **Input imagery** layer, click the **Layer** button and select the **Alexandra_OrthoMosaic_90cm** layer.

2. For **Processing mode**, accept the **Process as a mosaicked image** default option.

3. In the **Model settings** section, for **Model for pixel classification**, click **Select model**.

4. In the **Select item** pane, click **My content** and click **Living Atlas**.

5. In the search box , type land cover dlpk and press Enter.

   The search results contain several deep learning package (dlpk) models for land-cover classification. Each model is designed for a different type of imagery input, such as Sentinel-2, Landsat 8, high-resolution satellite imagery, or aerial imagery.

6. Select **High Resolution Land Cover Classification - USA** and click **Confirm**.

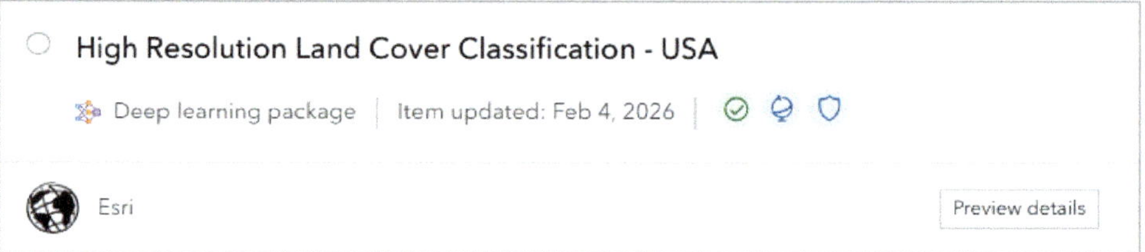

   This model is designed for high-resolution imagery, which matches the data you're working with. After a few moments, the model arguments appear in the tool.

7. Click the pop-out button for the deep learning model to see the item's page.

   This model is trained to work on 8-bit, 3-band imagery. The model was trained on imagery that has an 80 to 100 cm (or 0.8 to 1 m) resolution, which means that it will best perform with input of a similar resolution. That matches the 8-bit, 3-band, 90 cm imagery that you have for Alexandra.

8. When you have finished examining the model description information, close the item description tab.

9. In the **Model arguments** section, find the **Batch Size** argument.

   Deep learning pixel classification cannot be performed on the entire image at one time. Instead, the tool will split the image into small tiles, based on the Tile Size value. A Batch Size of 4 means that the tool will process four image tiles at a time. For this tutorial, you'll keep the default value of 4, and you'll also accept the other argument default values.

10. In the **Result layer** section, for the **Output name** parameter, type Land_Cover_Raster and add your name or initials.

11. Click **Environment settings** to expand the section.

    You'll accept the default values for the coordinate system, transformation, and extent. When you are testing a deep learning tool on your own imagery, you can set a processing extent to process a subset of your image and test the output. This reduces credit consumption and allows you to assess the tool settings.

12. In the **Cell size** section, click **Select cell size** and click **As specified**.

13. In the resulting **Cell size** value box, type 0.9.

    As mentioned earlier, the model is expecting input imagery with an 80–100 cm cell size (or 0.8–1 m). The imagery for Alexandra has been resampled to that larger cell size before being used as input for deep learning classification. The resampled imagery will be much closer to the model's expectation. This will ensure a faster process and more accurate land-cover classification results.

14. Click **Estimate credits**.

    This operation should consume about one credit.

15. Click **Run**.

    This process may take 10–15 minutes.

16. At the upper right of the tool pane, click the **History** button to view the processing status.

17 When the tool is finished running, view the result on the map.

The resulting layer is a raster imagery layer in which each pixel value indicates one of nine land-cover categories.

Examine the land-cover results

1. If the **Properties** pane is not visible, on the **Settings** toolbar, click the **Properties** button.

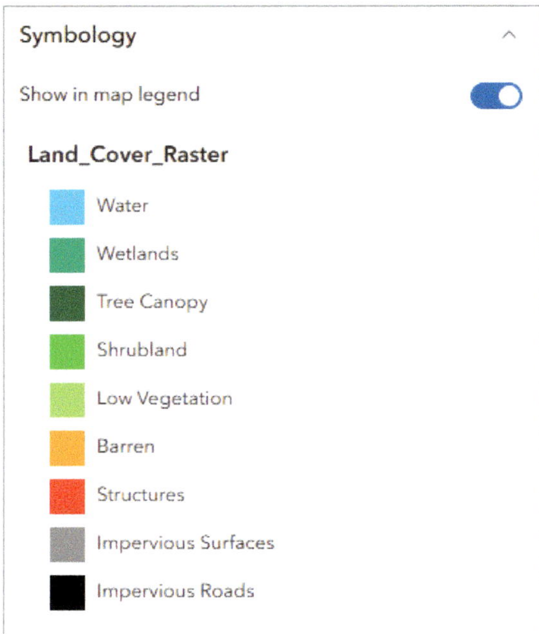

The Symbology section of the Properties pane shows the classes of the raster, with different colors representing the land-cover categories.

2. Pan and zoom around the map to explore the results.

Turning the layer off and viewing an imagery basemap below the new layer will help you evaluate the classified result.

3. On the **Contents** toolbar, click **Basemap** and select the **Imagery** basemap.

4. Return to the **Layers** pane. For the **Land_Cover_Raster** layer, click the visibility button to turn the layer on and off as you explore the results.

You may notice that the vegetation areas were extracted with good overall accuracy, but the building classification is somewhat lower quality, especially in the areas with informal housing. Depending on the imagery resolution and the specific types of buildings present, the quality of results may vary.

5. Save your map.

Take the next step (optional)

Now that you have a land-cover raster, you can derive a polygon layer from it. This will allow you to conduct analyses using vector-based geoprocessing tools available in ArcGIS Online. You can do that with the Convert Raster to Feature tool. The output will be a hosted feature layer in your ArcGIS Online content.

1. Search for and open the **Convert Raster to Feature** tool. Apply the following settings:
    - For **Input layer**, select **Land_Cover_Raster**.
    - For **Field to convert**, select **Class**.
    - For **Output type**, select **Polygon**.
    - Uncheck the box for **Simplify lines or polygons**. This preserves the original detail of the land cover in the output features for more accurate results.
    - Set the **Output features name** to Land_Cover_Feature and add your name or initials.

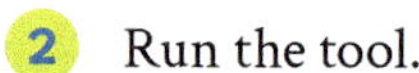

2. Run the tool.

This layer is now ready to be styled and added to your workflow.

Apply this workflow to your own imagery

To apply this workflow to your own imagery, keep the following in mind:

Storing your imagery: In this tutorial, you used an image layer that was generated in Site Scan for ArcGIS out of raw drone imagery, resampled in ArcGIS Pro, and saved to ArcGIS Online. When working with your own data, you can upload it directly to ArcGIS Online or copy it to your local computer and upload it to ArcGIS Online as a tiled imagery layer.

Preparing your imagery: The three bands expected are red, green, and blue (or RGB). If your imagery has more than three bands, you should extract the relevant bands before proceeding with the deep learning process. The model also expects the imagery to have an 8-bit pixel depth. If your imagery has a different pixel depth, such as 16 bit, you should convert it to 8 bit.

Finding information about your imagery: If you are not sure what your imagery's properties are (such as number of bands, pixel depth, or cell size), click your imagery layer and click **Properties**. In the **Information** section of the **Properties** pane, click the pop-out button to see the ArcGIS Online item description page, and under **Image properties**, find the number of bands and cell size values.

Experimenting with the cell size: As you are using the **Classify Pixels Using Deep Learning** geoprocessing tool, you can try several cell size values to see which one gives you the best result for your imagery. However, 0.9 should be a good fit for this model, since it is expecting an 80–100 cm cell size (or 0.8–1 m). It is best not to use imagery that has a coarser cell size than what the model expects.

Experimenting on a small extent: While experimenting, you can limit the processing to a small extent for faster results. In the **Environments** section of the tool, under **Processing Extent**, you can choose **Full extent**, **Coordinates**, **Display extent**, or **Layer**. It can be useful to zoom to a subset of the image and process only the display extent.

Summary

In this chapter, you used ArcGIS Online to run a pretrained deep learning model that classified high-resolution drone imagery into a raster layer showing the land cover types in Alexandra, South Africa.

This chapter is based on an Esri tutorial by Lisa Tanh.

CHAPTER 9

Training a regression model to estimate biomass in aerial imagery

Objectives

- Prepare explanatory variables using GEDI and Landsat 9 imagery.
- Train and review a regression model.
- Run a prediction tool.

Introduction

Conventional biomass measurement involves significant labor and time, but advances in deep learning technology have made it possible to produce high-quality estimates using remotely sensed data. In this chapter, you'll train a regression model and use it to predict biomass estimates in ArcGIS Pro.

You'll set up the ArcGIS Pro project and examine the input data. You'll review sample data and prepare seven spectral indexes derived from a Landsat 9 scene and a digital elevation model (DEM) raster to be used as explanatory variables. Next, you'll use all this data to train a regression model and capture the relationships between known Aboveground Biomass Density (AGBD) values and explanatory variables. You will then examine the performance of your model, do some data cleanup, and retrain your model to obtain higher performance. Then, you'll use the resulting model to predict AGBD values throughout the study area.

Requirements

- ArcGIS Pro
- ArcGIS Image Analyst

Tutorial 9-1

Review and prepare variables

Set up the project

To predict the aboveground biomass (AGB) throughout several counties in Georgia, you'll use the following data:

- Target sample data: This will be a set of known AGB values for sample locations. You'll use point data extracted from a Global Ecosystem Dynamics Investigation (GEDI) satellite lidar trajectory dataset.
- Explanatory variables: This will be data that can explain the AGB sample values and can then help predict AGB values for new areas. You'll use Landsat 9 multispectral satellite imagery, a digital elevation model (DEM), and additional derived raster layers.

The Landsat 9 multispectral satellite imagery was chosen because the sensor's spectral characteristics respond to vegetation, which is directly related to biomass. A DEM captures the topological variability and terrain complexity, which can also influence vegetation growth.

During training, the model will capture the relationships between sample values and explanatory variables. Once you are satisfied with the model, you'll use it to create a raster that predicts AGB values throughout the extent.

1 To get started, open **C:/GeoAI_Data** and open the **Chapter09** folder. Double-click **Estimate_Biomass.aprx** to open the project in ArcGIS Pro.

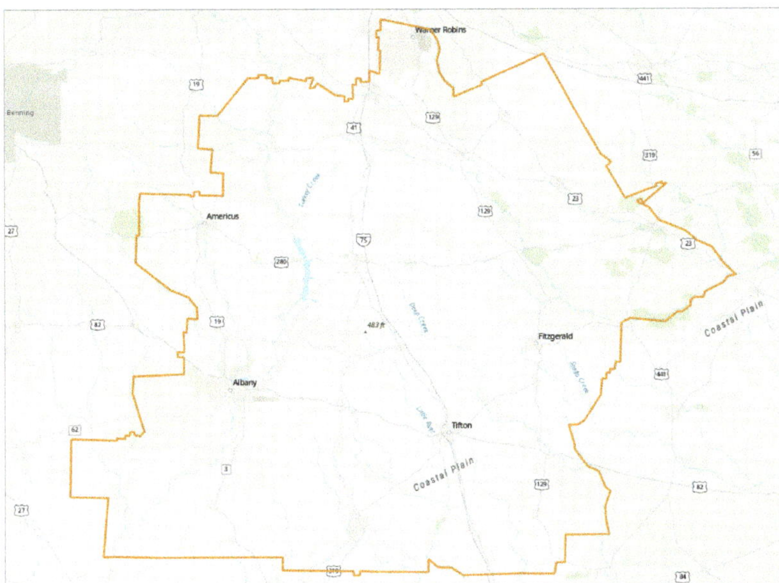

The map displays the study area boundary as a polygon outlined in orange. This area represents 20 counties in Georgia.

2 In the **Contents** pane, check the box next to the **Landsat9** layer to turn it on.

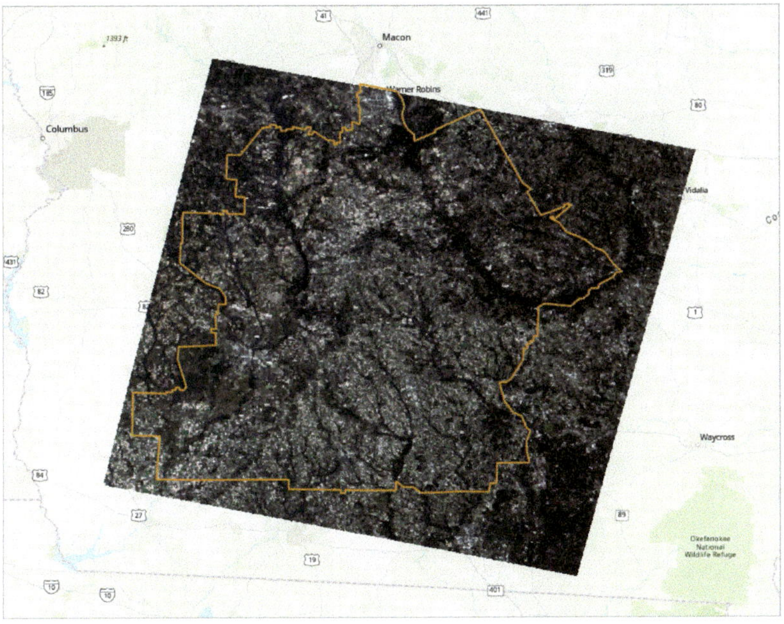

This is a Landsat 9 satellite imagery scene that includes seven spectral bands with surface reflectance values:

- Band 1—coastal/aerosol
- Band 2—blue
- Band 3—green
- Band 4—red
- Band 5—near-infrared (NIR)
- Band 6—shortwave infrared (SWIR) 1
- Band 7—shortwave infrared (SWIR) 2

These bands will be used as explanatory variables. The image rendering is a combination of the red, green, and blue bands, which shows colors close to what the human eye would usually see.

Next, you'll review the DEM in the map.

Examine the data

1. In the **Contents** pane, turn on the **DEM.tif** layer and review it.

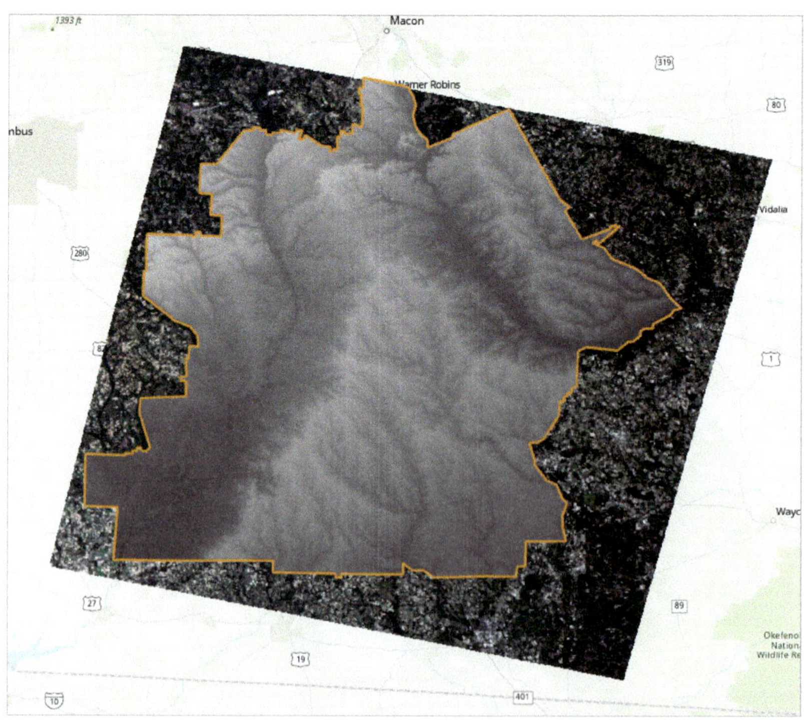

This layer will also be used as an explanatory variable. Next, you'll review the GEDI data.

2. Turn off visibility for the **DEM** and **Landsat9** layers since you won't need them for the next steps.

3. In the **Catalog** pane, expand **Folders** > **Chapter09** > **InputData** > **GEDI_L4A**.

This folder contains eight GEDI files that will be used as the samples with known AGB values, or training targets. Note that these are trajectory HDF5 files; they are not raster files but trajectory data.

AGB represents living vegetation above the ground, measured as mass per unit, typically megagram (that is, metric ton) per hectare. Measuring AGB physically on the ground over a large study area is labor intensive and nearly impossible. Estimating AGB using remote sensing data is a good alternative solution.

GEDI is a satellite lidar mission from NASA that measures the 3D structure of the Earth's surface, including canopy height—made up of the trees and shrubs that constitute biomass. GEDI captures sample points along the sensor's tracks. From those measurements, the AGBD can be derived. The GEDI L4A product contains these derived AGBD point values. This data is delivered as trajectory-structured HDF5 files and can be brought into ArcGIS as a trajectory dataset.

A trajectory dataset with the relevant AGBD data has been prepared for you, as shown in the AGBD_observations layer. It will be used for training later in the workflow.

4. In the **Contents** pane, turn on the **Gedi** layer. Turn the **Footprint** and **Point** sublayers on and off to familiarize yourself with the full trajectory dataset that contains all the tracks and points. Zoom in to an area of your choice until you see the individual points.

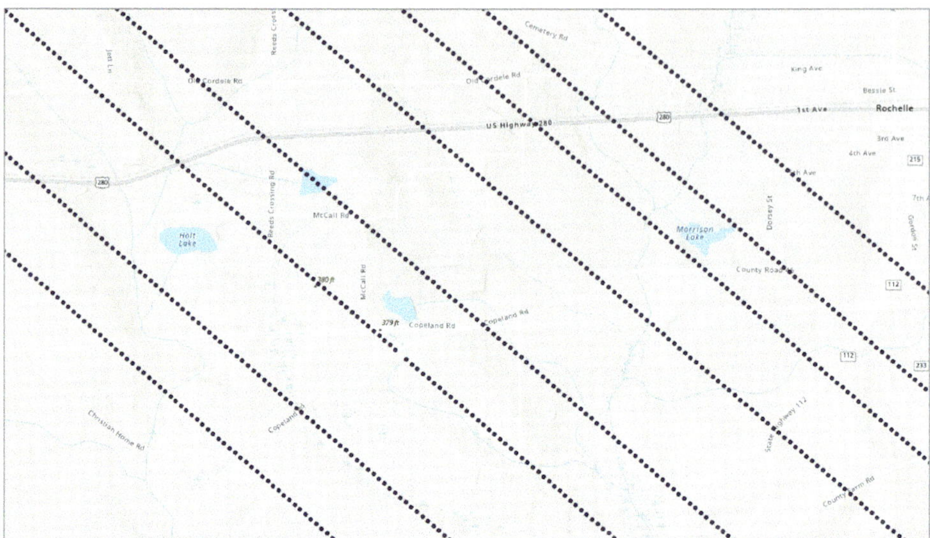

Each point contains an AGBD value.

Extract the relevant AGBD point data

Only the GEDI points within the study area are relevant to your workflow. These have been prepared in the AGBD-observations layer and have been clipped to the area of interest (AOI) so you don't need the full dataset.

1. In the **Contents** pane, right-click the **Gedi** layer and click **Remove** since you won't need it for the rest of this workflow.

2. Turn on the **AGBD_observations** layer. Right-click it and click **Zoom To Layer**.

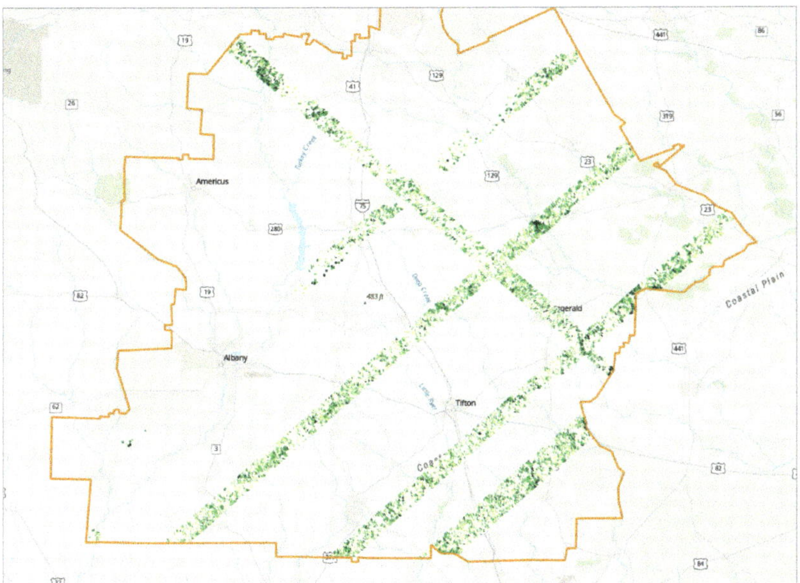

You can see that the AGBD_observations layer contains only the points within the study area. The layer displays with a symbology where the points in dark-green tones indicate the highest AGBD values and the points in light-yellow tones indicate the lowest AGBD values. This layer will be used as known samples, or training targets, during the model training.

3. In the **Contents** pane, right-click the **AGBD_observations** layer and click **Attribute Table**.

The AGBD_observations attribute table appears. Each row corresponds to a point, and the AGBD field gives the aboveground biomass density value for each point (in metric tons per hectare). In total, this layer has 106,159 points.

	OBJECTID *	SHAPE *	lon	lat	Time	AGBD	TrajectoryID
1	1	Point	-83.603913	31.035486	9/24/2022 8:43:44.527 PM	234.650955	1
2	2	Point	-83.603486	31.035865	9/24/2022 8:43:44.535 PM	136.511093	1
3	3	Point	-83.603082	31.036228	9/24/2022 8:43:44.543 PM	39.995003	1
4	4	Point	-83.602668	31.036598	9/24/2022 8:43:44.552 PM	49.484825	1
5	5	Point	-83.602249	31.036972	9/24/2022 8:43:44.560 PM	17.345238	1
6	6	Point	-83.601838	31.037339	9/24/2022 8:43:44.568 PM	32.382404	1
7	7	Point	-83.601422	31.03771	9/24/2022 8:43:44.577 PM	52.036236	1
8	8	Point	-83.601008	31.038081	9/24/2022 8:43:44.585 PM	142.240097	1
9	9	Point	-83.600594	31.03845	9/24/2022 8:43:44.593 PM	148.676041	1
10	10	Point	-83.600178	31.038822	9/24/2022 8:43:44.601 PM	307.7771	1
11	11	Point	-83.599763	31.039192	9/24/2022 8:43:44.610 PM	216.754684	1
12	12	Point	-83.599333	31.039574	9/24/2022 8:43:44.618 PM	31.818235	1
13	13	Point	-83.598924	31.03994	9/24/2022 8:43:44.626 PM	71.357536	1

0 of 106,159 selected

4 Close the attribute table and turn off the **AGBD_observations** layer.

Prepare Landsat 9–derived explanatory variables

You will create seven spectral indexes derived from the Landsat 9 scene to be used as explanatory variables. A spectral index combines spectral bands through a mathematical formula, usually computing some type of ratio. The resulting output is a new raster image that emphasizes a specific phenomenon, such as vegetation, water, urban development, or moisture. These spectral index layers will provide additional information to account for different vegetation conditions, in turn helping better predict AGB values.

You'll create several indices that will serve as additional explanatory variables:

- NDVI—normalized difference vegetation index
- EVI—enhanced vegetation index
- PVI—perpendicular vegetation index
- NBR—normalized burn ratio
- NDWI—normalized difference water index
- NDBI—normalized difference built-up index
- MSI—moisture stress index

You'll start with NDVI, used to differentiate healthy vegetation from unhealthy vegetation or absence of vegetation. You'll use the **Band Arithmetic** raster function.

1. On the ribbon, click the **Imagery** tab. In the **Analysis** group, click the **Raster Functions** button.

2. In the **Raster Functions** pane, in the search box, type Band Arithmetic and click the tool in the results.

3. In the **Band Arithmetic Properties** tool, apply the following settings:

   - For **Raster**, select **Landsat9**.
   - For **Method**, select **NDVI**.
   - For **Band Indexes**, type 5 4, corresponding to the near-infrared and red bands that are needed for the NDVI calculation. Ensure that you put a space between the numbers.

4. In the tool, click the **General** tab. For **Name**, type NDVI. Click **Create new layer**.

   A new layer named NDVI_Landsat9 is added to the map. The raster in the map contains calculated NDVI values ranging between –1 (absence of vegetation) and 1 (healthy vegetation).

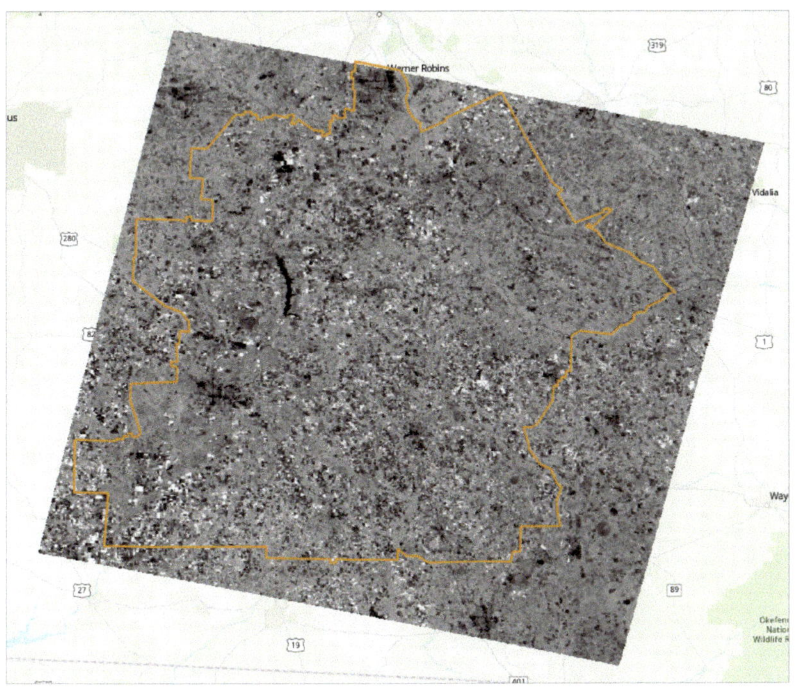

Next, you'll create the remaining spectral index layers—EVI, NBR, PVI, NDWI, and NDBI—following the same steps.

5 Repeat the previous steps with the following band settings:

Method/name	Description (for reference)	Band indexes	Band names (for reference)
EVI	Enhanced vegetation index	5 4 2	NIR, Red, Blue
NBR	Normalized burn ratio (used to identify burn scars)	5 7	NIR, SWIR 2
PVI	Perpendicular vegetation index	5 4 0.3 0.5	NIR, Red, a, b (slope and gradient values)
NDWI	Normalized difference water index	5 3	NIR, Green
NDBI	Normalized difference built-up index	6 5	SWIR 1, NIR

For MSI (moisture stress index), the Band Arithmetic raster function doesn't include the MSI option under Method, so you'll use the User Defined option to calculate it, spelling out the mathematical formula explicitly: B6 / B5, where the bands are referred to by B + [a band number]. So, this formula means that the SWIR 1 band should be divided by the NIR band.

6 In the **Band Arithmetic Properties** tool, apply the following settings:

- For **Raster**, select **Landsat9**.
- For **Method**, select **User Defined**.
- For **Band Indexes**, type B6 / B5.
- Under **General**, for **Name**, type MSI.

At the end of this process, all seven index layers should be added to the map and listed in the Contents pane.

Prepare DEM-derived explanatory variables

You will now derive an aspect layer from the DEM layer using the Aspect raster function. The aspect indicates the direction that each slope faces (north, south, east, west). It is relevant as an explanatory variable because solar illumination will vary according to the aspect value, and this will affect vegetation growth.

1. In the **Raster Functions** pane, search for and open the **Aspect** raster function.
2. In the **Aspect** raster function pane, for **Raster**, select the **DEM.tif** layer.
3. Click **Create new layer**.

   A layer named Aspect_DEM.tif is added to the map.

In the next section, you will use all the explanatory variable layers you created as input to the machine learning model.

Tutorial 9-2

Train a regression model

Next, you'll train your regression model and capture the relationships between known AGBD values and explanatory variables. You'll train the model to predict biomass with the Train Random Trees Regression Model tool.

Run the regression model training tool

1. On the ribbon, click the **Analysis** tab. In the **Tools** group, click **Geoprocessing**.
2. In the **Geoprocessing** pane, search for and open the **Train Random Trees Regression Model** tool.
3. For **Input Rasters**, add **Landsat9**, **DEM.tif**, and all eight derived explanatory variable layers.

Caution: Use the exact same order for these layers now in the **Train Random Trees Regression Model** tool and later in the **Predict Using Regression Model** tool.

4. For **Target Raster or Points**, select **AGBD_observations**.
5. For **Target Value Field**, select **AGBD**.

The resulting output model will be an .ecd file. You'll choose a name for it.

6. For **Output Regression Definition File**, click the browse button.

7. In the **Output Regression Definition File** window, browse to **Folders** > **Chapter09**, and for **Name**, type Biomass_model.ecd and click **Save**.

   The output will also include some auxiliary files that you can use to understand the model's accuracy. You'll set up their names.

8. In the tool pane, expand **Additional Outputs**.

9. For **Output Importance Table**, click the browse button, browse to **Folders** > **Chapter09**, and for **Name**, type Importance.csv. Click **OK**.

10. For **Output Scatter Plots**, click the browse button, browse to **Folders** > **Chapter09**, and for **Name**, type Biomass_scatterplots.pdf. Click **Save**.

    Finally, you will also set up the training option parameters.

11. Expand **Training Options**.

12. For **Percent of Samples for Testing**, type 5, and accept the other default values.

    Note: The 5 percent value (instead of the default 10 percent) ensures that less data will be set aside for testing and more will remain available for training.

13. Click **Run**.

    After a couple of minutes, the model training is complete.

Review the model performance

To understand the model performance, you will now review the outputs from the Train Random Trees Regression Model tool. Machine learning workflows are often iterative. You must decide whether the model is performing optimally or whether cleaning up some of the input data could improve its performance. In the latter case, you will need to retrain the model using the cleaned-up data.

First, you will look at the content of the Importance.csv table, which shows how each explanatory variable contributed to predict the target sample values. You'll create a bar chart to summarize that information.

1. In the **Contents** pane, under **Standalone Tables**, right-click the **Importance.csv** table layer. Click **Create Chart** > **Bar Chart**.

An Importance.csv chart pane and a Chart Properties pane appear.

2. In the **Chart Properties** pane, apply the following settings:

- For **Category or Date**, select **Explanatory_Variables**.
- For **Aggregation**, select **None**.
- Under **Numeric field(s)**, click **Select**, check the box for **Importance**, and click **Apply**.

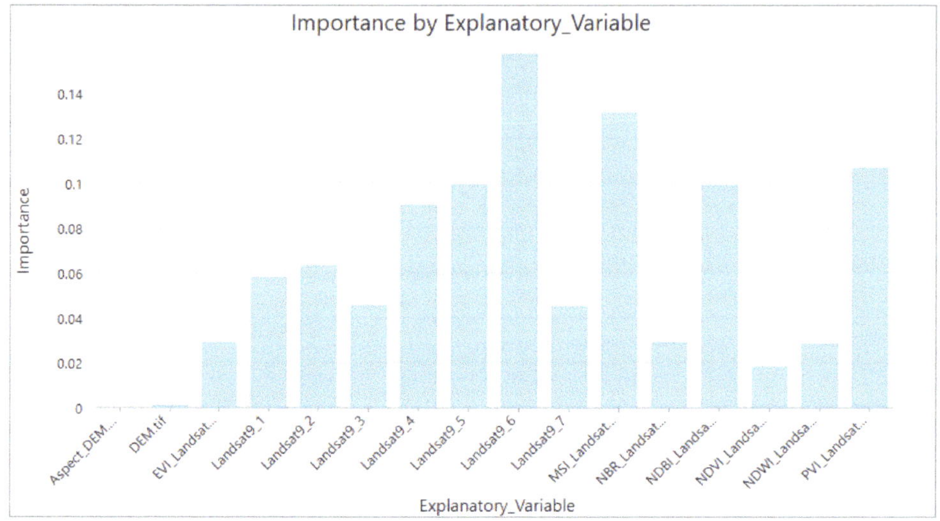

In the Importance.csv chart pane, the Importance by Explanatory_Variable chart appears.

You can observe that the Landsat spectral bands, especially SWIR 1 (Landsat9_6) and near-infrared (Landsat9_5), play important roles in explaining (or predicting) the biomass values. Additionally, several band indexes make substantial contributions, especially MSI_Landsat9, PVI_Landsat9, and NDBI_Landsat9. On the other hand, the DEM and Aspect_DEM layers contribute the least, which makes sense because this study area is mostly flat terrain. However, in other extents with more elevation variation, the importance of the elevation data would probably be higher. Next, you'll review the scatterplots document.

Note: The Random Trees algorithm is not deterministic, so the results you obtain may vary slightly.

3 Close the **Importance.csv** chart pane.

4 In File Explorer, browse to **C:/GeoAI_Data/Chapter09** and double-click the **Biomass_scatterplot.pdf** file to open it.

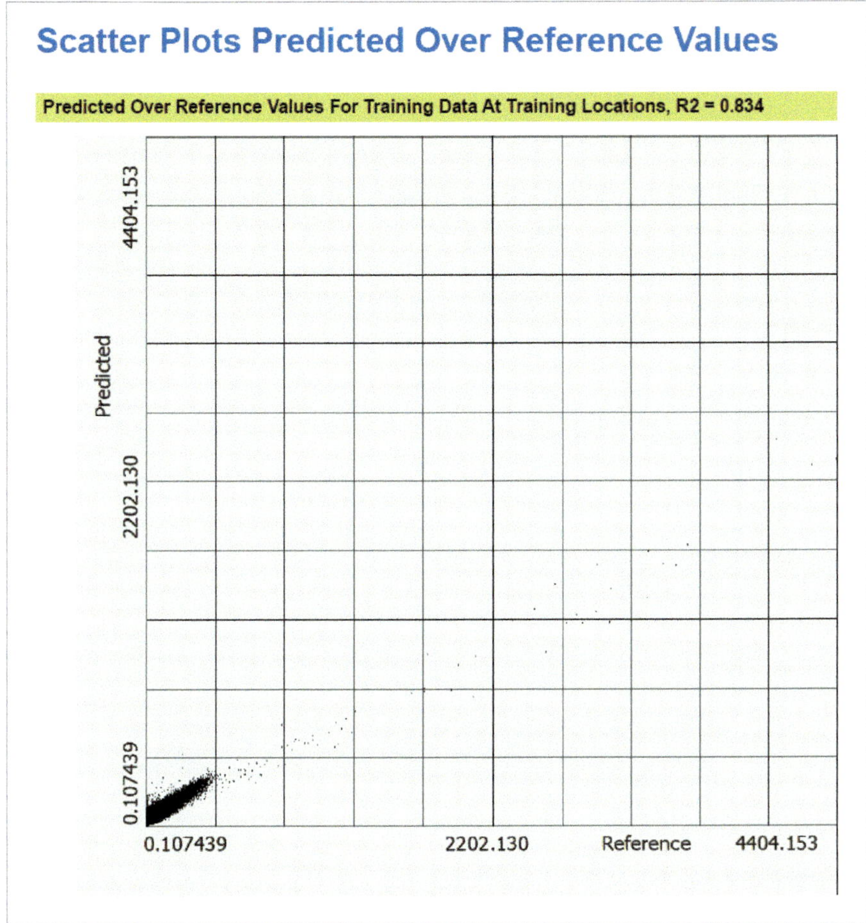

In the PDF, the first scatterplot shows the following for each sample point used in training:

- The original known value (x-axis)
- The predicted value, after the training is complete (y-axis)
- The R^2 value, ranging from 0 to 1, serves as an indicator of the model's performance. An R^2 value of 0.834 for the training performance is acceptable. However, although most values are concentrated under 1,000, you can observe some extremely high values scattered from a bit under 1,000 to over 4,000.

Since these points might be erroneous outliers that degrade the model's learning performance, you will need to evaluate whether you should keep these extreme points or remove them from the training data. First, you will look at a histogram chart for the AGBD_observations layer to choose a more precise threshold for the outlier points.

5 Close the PDF and switch back to ArcGIS Pro.

6 In the **Contents** pane, open the attribute table for **AGBD_observations**.

7 In the attribute table, right-click the **AGBD** field heading and select **Visualize Statistics**.

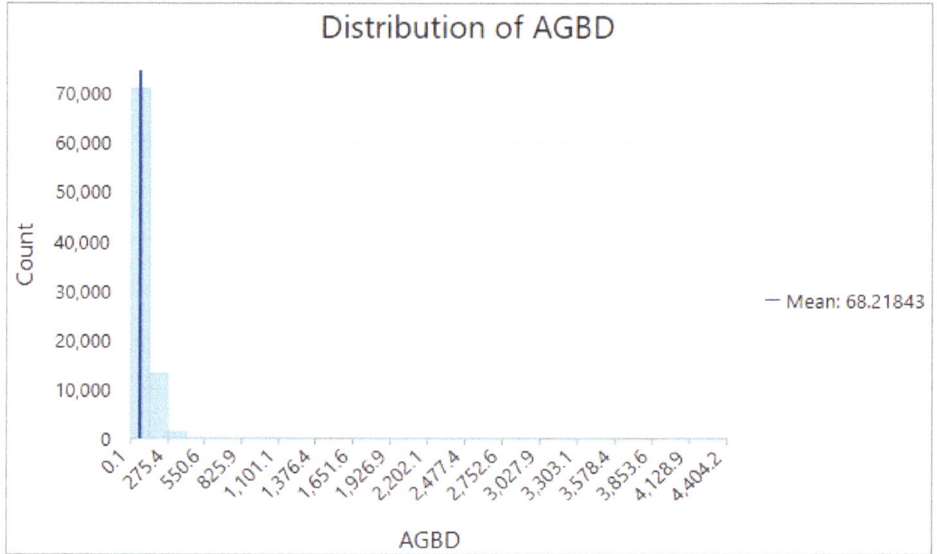

The statistics for the AGBD field appear in a histogram chart named Distribution of AGBD.

You can see that most of the points have AGBD values that are less than 700, with only a few points having values greater than 1,000. You will choose 1,000 as the threshold to define outlier points.

8 Close the chart and attribute table.

You will now modify the display on the map to make the exploration of the high-value points easier.

9 In the **Contents** pane, drag the **Landsat9** layer to position it just above **Aspect_DEM.tiff**.

10 Turn on the **AGBD_observations** and **Landsat9** layers.

11 Right-click the **AGBD_observations** layer and click **Symbology**. Change the **Primary symbology** to **Single Symbol**.

This symbology will make it easier to see the points you select on the map. You will now select the high-value AGBD points.

12 In the **Contents** pane, select the **AGBD_observations** layer.

13 On the ribbon, on the **Map** tab, in the **Selection** group, click **Select By Attributes**.

14 In the **Select By Attributes** window, under **Expression**, create the expression `Where AGBD is greater than 1000` and click **OK**.

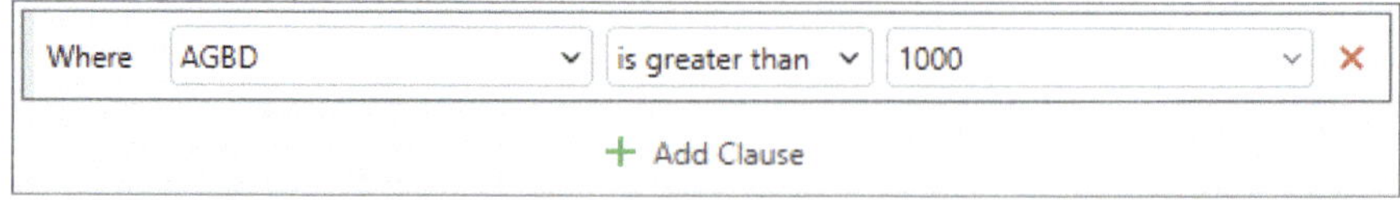

About 40 points are selected on the map.

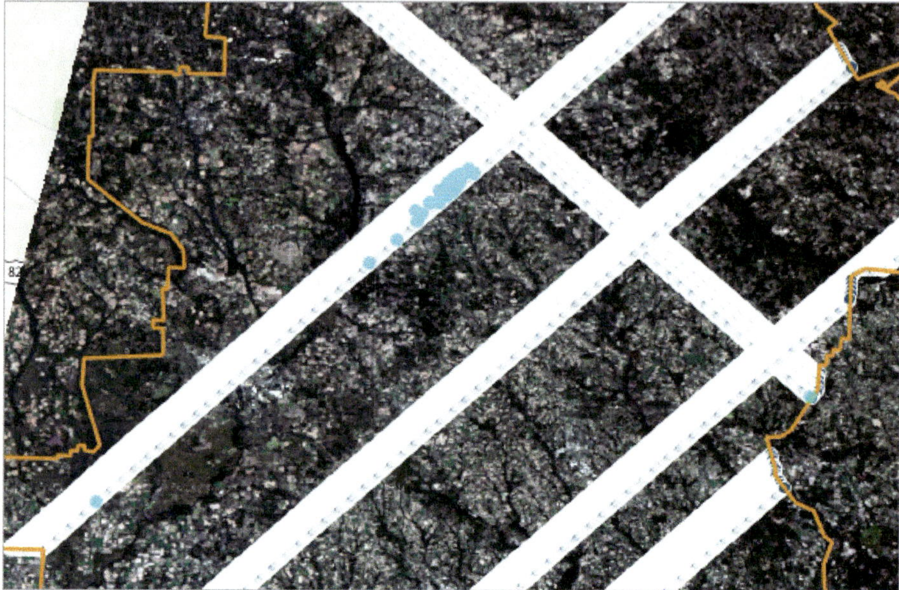

You will now review a few of these points individually.

15 Open the **AGBD_observations** attribute table and click the **Show Selected Records** button at the bottom of the pane.

Only the selected features are now listed in the table.

16 Double-click the row header for the first feature.

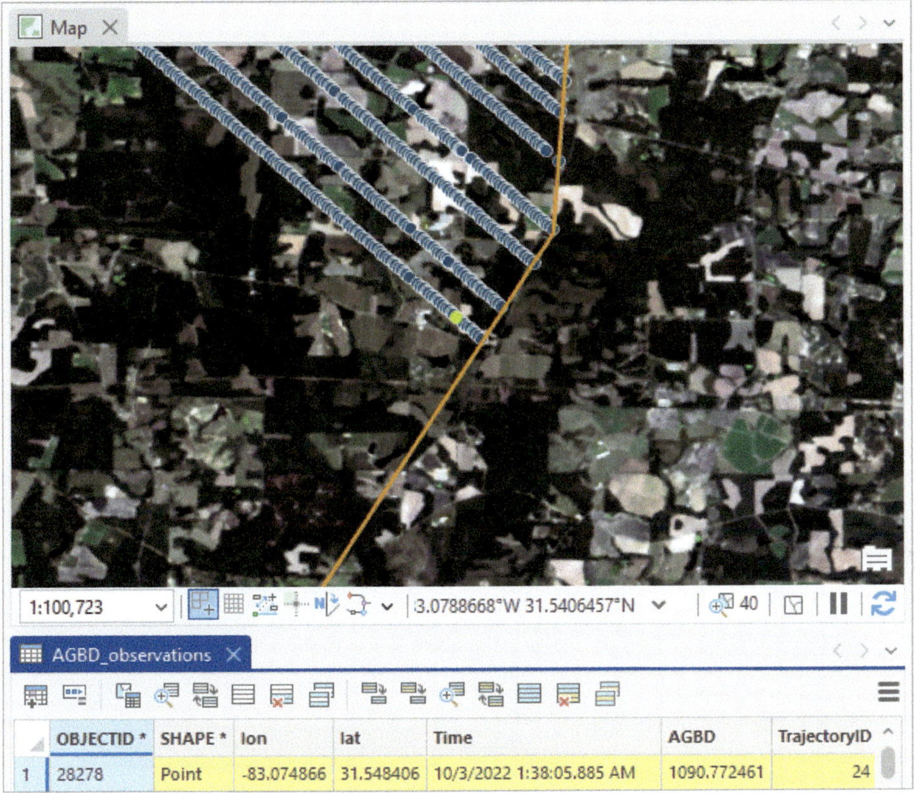

On the map, the point is zoomed to and highlighted.

17 Zoom in until you can see the imagery details underneath.

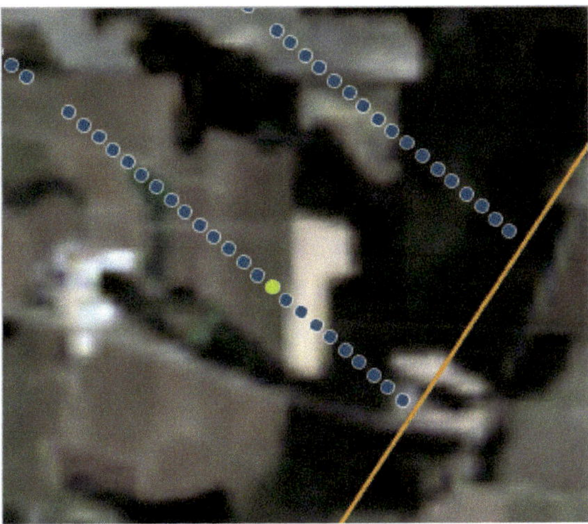

The point falls on some type of not-so-dense grass field, which should not have an AGBD value above 1,000. In contrast, you can see that neighboring points don't appear in cyan, because they were not selected. This means their AGBD value is under 1,000 and they don't have a similar, abnormally high value.

18 In the attribute table, double-click the row header for the third feature.

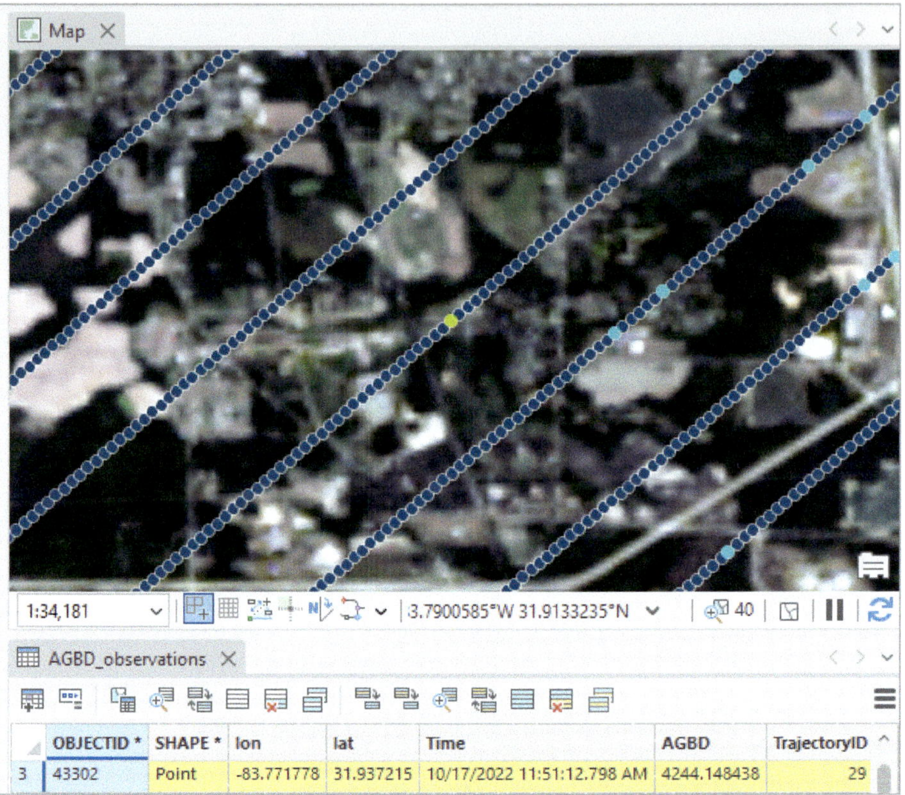

This point also falls on some type of grass field, which should not have a value above 1,000. You can see that these high-value points are outliers that must be faulty. You will delete them.

Clean up AGBD observations and retrain the model

You'll now delete the high-value outlier points. You'll also delete the points that have a null value, since they are of no use for training. Then, you'll retrain the model.

1. In the **Contents** pane, right-click **AGBD_observations** and click **Zoom To Layer**.
2. Open the **Select By Attributes** window and confirm the first clause **Where AGBD is greater than 1000** is still present.

   You will add a second clause to select the features with null values.

3. Click the **Add Clause** button. Then, create the following expression: `Or AGBD is null`. Click **OK**.

   In the AGBD_observations attribute table, there are now more than 20,000 points selected that have either abnormally high values or null values.

4. On the attribute table toolbar, click the **Delete** button.
5. If prompted to confirm that you want to delete the data, click **Yes**.

   You will save these edits.

6. On the ribbon, click the **Edit** tab. In the **Manage Edits** group, click **Save**.

   The selected points are deleted from the AGBD_observations feature class. Next, you will rerun the training tool with the updated data to obtain a higher-performing model.

7. On the ribbon, on the **Analysis** tab, in the **Geoprocessing** group, click **History**.

   The History pane appears. It contains the history of all the tools you have run in this project.

8. In the **History** pane, double-click the **Train Random Trees Regression Model** entry.

   The Train Random Trees Regression Model tool appears, with all the parameter values you used originally. You will rename the outputs so that they don't overwrite the original results.

9 For **Output Regression Definition File**, change the name to Biomass_model2.ecd.

10 Expand **Additional Outputs**. Rename the **Output Importance Table** to Importance2.csv and rename the **Output Scatter Plots** to Biomass_scatterplots2.pdf.

11 Click **Run**.

After a couple of minutes, the model is retrained.

12 In File Explorer, browse to **C:/GeoAI_Date/Chapter09** and double-click the **Biomass_scatterplots2.pdf** file to open it.

In the PDF, in the first scatterplot, you can see that the model performance has improved to a $R^2 = 0.888$ (up from $R^2 = 0.834$ previously). You can also note that all the values in the plot are now lower than 1,000.

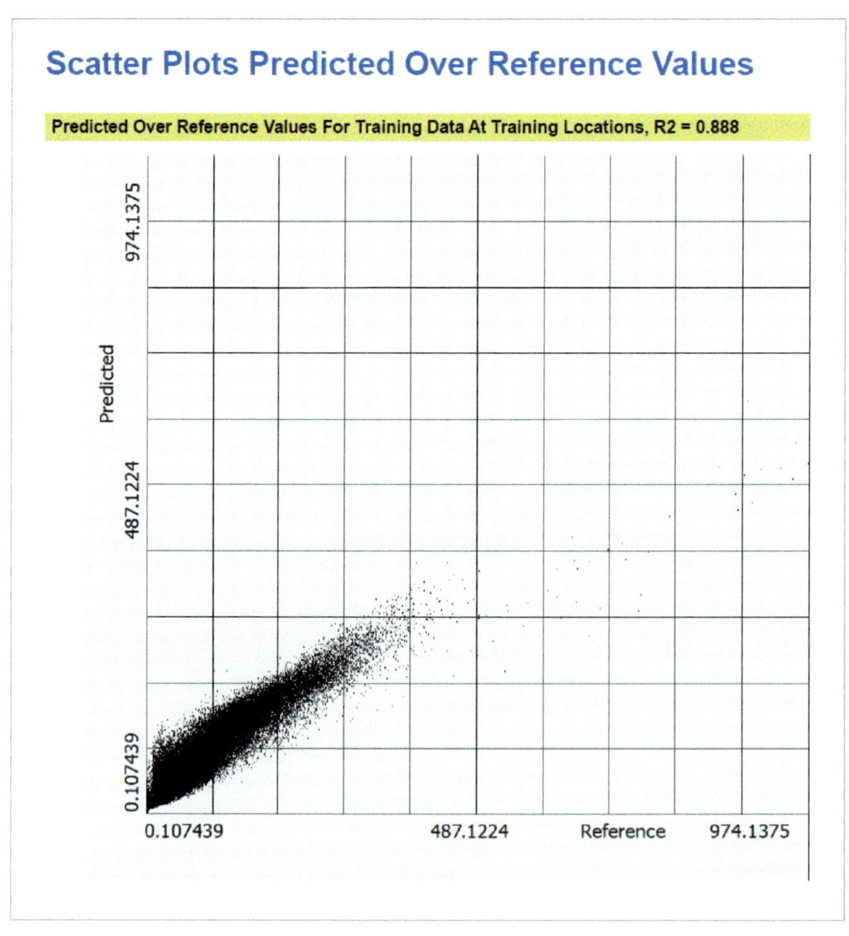

You have also obtained better results in the second and third scatterplots in the PDF, which show the model performance on test points.

13 Close the PDF and switch back to ArcGIS Pro.

Tutorial 9-3

Predict biomass using the trained model

You'll now use the model to predict biomass for the entire study area. You will do that with the Predict Using Regression Model tool. The input will use the same explanatory variables that you used for the model training.

Run the prediction tool

1 In the **Geoprocessing** pane, search for and open the **Predict Using Regression Model** tool. Apply the following settings:

- For **Input Rasters**, add **Landsat9**, **DEM**, and all eight derived layers in the same order as before.

> **Caution:** You must use the same order for these layers as you did earlier in the **Train Random Trees Regression Model** tool.

- For **Input Regression Definition File**, click the browse button, browse to **Folders** > **Chapter09**, click **Biomass_model2.ecd**, and click **OK**.
- For **Output predicted raster**, type Biomass_prediction.crf.

2 Click **Run**.

After a few minutes, the resulting layer is added to the map. You will now symbolize it.

3 In the **Contents** pane, right-click **Biomass_prediction.crf** and click **Symbology**.

4 In the **Symbology** pane, click **Color scheme** to open the list of colors.

5 At the bottom of the list, check the box for **Show names**. Then, scroll through the list to find the **Blue-Green (Continuous)** color scheme.

6 Turn off all the layers except the **Biomass_prediction.crf**, **AOI**, and basemap layers.

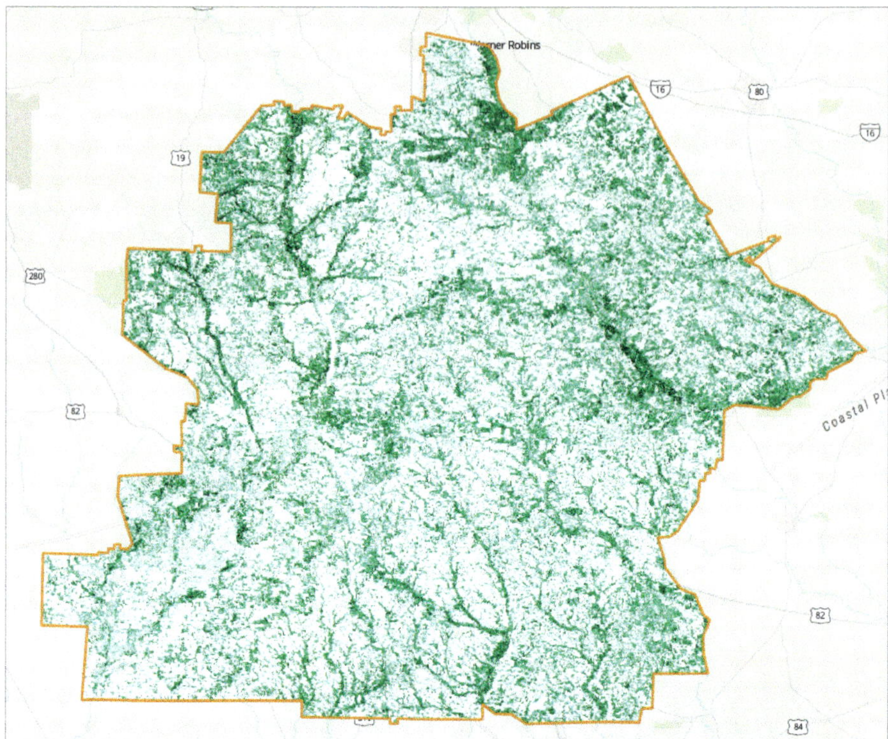

Dark-green tones indicate the areas with the highest biomass density, and light or white tones indicate low density or absence of biomass.

> **Note:** To apply a similar workflow to large areas that are represented across several Landsat scenes, it is recommended that you first address cloud and shadow removal and compose these images in a mosaic dataset and consider using the capabilities of direct data access and cloud-based computing using ArcGIS Pro.

Take the next step (optional)

If you wanted to compute the biomass density per county, you could use the Counties polygon layer and the Zonal Statistics as Table tool to find the average biomass density per county and generate a chart to give an overview of your results.

1. In the **Zonal Statistics as Table (Image Analyst Tools)** tool pane, apply the following settings and run the tool:
   - For **Input Raster** or **Feature Zone Data**, select **Counties.**
   - For **Zone Field**, verify that **Name** is selected.
   - For **Input Value Raster**, select **Biomass_prediction.crf**.
   - For **Output Table**, type Average_biomass_by_county.
   - For **Statistics Type**, select **Mean**.
2. In the **Contents** pane, under **Standalone Tables**, right-click the **Average_biomass_by_county** table, click **Create Chart**, and choose **Bar Chart**.
3. In the **Chart Properties** pane, on the **Data** tab, apply the following settings:
   - For **Category or Date**, select **NAME**.
   - For **Aggregation**, select **<none>**.
   - Under **Numeric field(s)**, click **Select**, check the box for **MEAN**, and click **Apply**.
   - Under **Sort**, select **Y-axis Descending**.
4. Click the **General** tab and apply the following settings:
   - For **Chart title**, type Average biomass by county.
   - For **X axis title**, type Counties.
   - For **Y axis title**, type Biomass density (in metric tons per hectare).
5. In the **Average_biomass_by_county** chart pane, view the **Average biomass by county** chart.

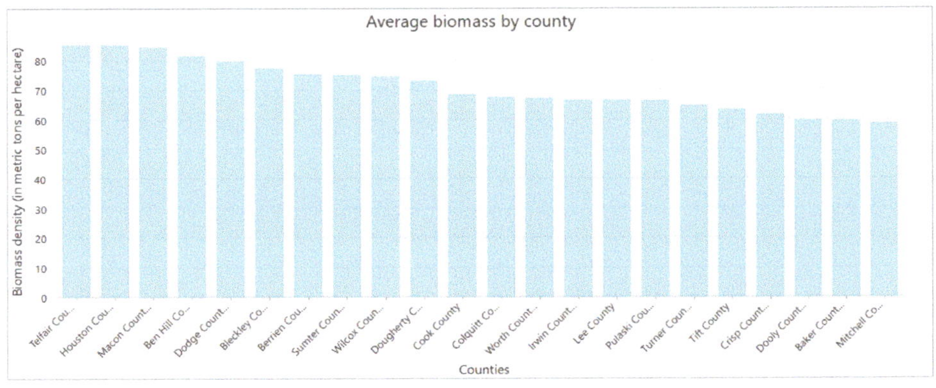

You can also join the Biomass_by_county table to the Counties layer to create a thematic map showing the average biomass by county.

Summary

In this tutorial, after setting up the project and examining the data, you prepared a trajectory dataset containing GEDI data and extracted the relevant AGBD point data for the study area. You used raster functions to prepare explanatory variables. You then trained a model to predict biomass density. You examined the performance of the model, did some data cleanup, and retrained the model to obtain a higher performance. You used this better-performing model to predict biomass density throughout your study area. And finally, you summarized the results to obtain the average biomass density per county in the study area.

This chapter is based on an Esri tutorial by Hong Xu and Shengan Zhan.

Afterword

What's next

Now that you have hands-on experience with GeoAI and are likely contemplating the many uses for your exciting new skills, you may also be wondering, *What else can I explore?* Alongside the benefits of enhanced analysis that GeoAI brings, there is a growing ecosystem of complementary intelligence and reasoning technology to learn about. While GeoAI addresses challenges related to analysis, accuracy, and scale, Esri is introducing AI assistants throughout ArcGIS to enhance the user experience, hasten adoption, and maximize the value of investments already made in geospatial analysis.

Look for generative AI features that offer the capacity for consultation, expediting your investigation of geospatial science. AI capabilities run the gamut: AI assistants can prepare and interpret data, generate and refine analysis workflows, and perform mapping, visualization, and administrative tasks. This technology is helping scientists, analysts, and newcomers alike to focus their efforts where it matters most, on understanding and solving problems. Built on well-documented methods and trained with meaningful datasets, generative AI can aid your work with agentic applications and workflow-based actions. Its outputs are most valuable when guided by your intent and applied based on your professional judgment. AI technology can accelerate scientific exploration when paired with human oversight and ethical responsibility. Bottom line, you must ask the difficult questions and validate the analysis with context and care.

Systems-level thinking is required to adopt AI effectively. Intentional design means incorporating high-quality data, planning for interoperability across the geospatial ecosystem, and embracing agile approaches that allow you to add advances at a rapid pace and scale. As you grow, make room to adopt the ArcGIS Well-Architected Framework, which provides guidance on building stable, adaptable system architectures that allow for growth and innovation, incorporating AI where it can add value.

Remember that technology evolves along a continuum. For GIS, that evolution moved from command line to desktop, expanded to servers and shared data models, and leapt to the web and the cloud. AI is not merely a shiny new tool, it's another paradigm shift, expanding who can participate in advanced geospatial analysis and transforming how knowledge is created and shared. Our systems of record are changing into a system of understanding, with geography serving as the fundamental connection.

—Jennifer Weaklend
Principal business manager at Esri

Data credits

Chapter 2

Tuborg_Havn.tif: This 2019 aerial orthophoto comes from the Danish government website Styrelsen for Dataforsyning og Effektivisering.

World Topographic Map sources: Esri, TomTom, Garmin, FAO, NOAA, USGS, © OpenStreetMap contributors, and the GIS user community.

World Hillshade layer sources: Esri, Maxar, Airbus DS, USGS, NGA, NASA, CGIAR, N Robinson, NCEAS, NLS, OS, NMA, Geodatastyrelsen, Rijkswaterstaat, GSA, Geoland, FEMA, Intermap, and the GIS user community.

Chapter 3

Seattle_imagery.jp2: Aerial imagery from the US National Agriculture Imagery Program (NAIP) managed by the US Department of Agriculture. NAIP imagery covering the entire United States can be downloaded from the USGS EarthExplorer website.

World Topographic Map sources: Esri, TomTom, Garmin, FAO, NOAA, USGS, © OpenStreetMap contributors, and the GIS user community.

World Hillshade layer sources: Esri, Maxar, Airbus DS, USGS, NGA, NASA, CGIAR, N Robinson, NCEAS, NLS, OS, NMA, Geodatastyrelsen, Rijkswaterstaat, GSA, Geoland, FEMA, Intermap, and the GIS user community.

Chapter 4

Alexandra_Orthomosaic drone imagery provided by South Africa Flying Labs.

All other layers created by Rami Alouta, Colin Kelly, and Delphine Khanna.

World Topographic Map sources: Esri, TomTom, Garmin, FAO, NOAA, USGS, © OpenStreetMap contributors, and the GIS user community.

Chapter 5

Woolsey Fire imagery was used courtesy of USAA.

Building features were derived from Los Angeles County's Countywide Building Outlines (2017) dataset.

The training sample was derived from Los Angeles County's Countywide Building Outlines (2017) dataset and manually classified as Damaged or Undamaged.

Chapter 6

Sonoma lidar data: NASA Grant NNX13AP69G, the University of Maryland, and the Sonoma County Vegetation Mapping and LiDAR Program.

Topographic map sources: Esri, TomTom, Garmin, FAO, NOAA, USGS, © OpenStreetMap contributors, and the GIS user community.

Chapter 7

Sentinel-1 GRD imagery was acquired from the Copernicus Data Space Ecosystem.

Land Cover map 2018 classified raster was acquired from the European Union's Copernicus Land Monitoring Service dataset.

The ClassifiedLULC2024 raster was produced by Esri using the above datasets.

The ClassifiedSARLULC deep learning package was produced by Esri using the above datasets.

World Topographic Map was created by Esri, HERE, Garmin, FAO, NOAA, USGS, © OpenStreetMap contributors, and the GIS user community.

World Hillshade map was created by Esri, Airbus DS, USGS, NGA, NASA, CGIAR, N Robinson, NCEAS, NLS, OS, NMA, Geodatastyrelsen, Rijkswaterstaat, GSA, Geoland, FEMA, Intermap, and the GIS user community.

Chapter 8

Alexandra_Orthomosaic drone imagery provided by South Africa Flying Labs.

Topographic map sources: Esri, TomTom, Garmin, FAO, NOAA, USGS, © OpenStreetMap contributors, and the GIS user community.

Chapter 9

Landsat9 scene: The Landsat 9 mission is a partnership between USGS and NASA. The scene was downloaded from the USGS EarthExplorer website.

DEM raster: USGS 30 m resolution SRTM DEM downloaded from the NASA Earthdata website.

GEDI data: The Global Ecosystem Dynamics Investigation (GEDI) is a joint mission between NASA and the University of Maryland. The GEDI data was downloaded from the NASA Earthdata website.

Counties layer: Derived from the USA Counties Generalized Boundaries layer from ArcGIS Living Atlas of the World. Original data from the US Census Bureau.

World Topographic Map sources: Esri, TomTom, Garmin, FAO, NOAA, USGS, © OpenStreetMap contributors, and the GIS user community.

World Hillshade layer sources: Esri, Maxar, Airbus DS, USGS, NGA, NASA, CGIAR, N Robinson, NCEAS, NLS, OS, NMA, Geodatastyrelsen, Rijkswaterstaat, GSA, Geoland, FEMA, Intermap, and the GIS user community.

Reference paper: Demol, M., H. Verbeeck, B. Gielen, J. Armston, A. Burt, M. Disney, L. Duncanson, J. Hackenberg, D. Kükenbrink, A. Lau, P. Ploton, A. Sewdien, A. Stovall, S. M. Takoudjou, L. Volkova, C. Weston, V. Wortel, and K. Calders. 2022. "Estimating Forest Above-Ground Biomass with Terrestrial Laser Scanning: Current Status and Future Directions." *Methods in Ecology and Evolution* 13:1628–39. https://doi.org/10.1111/2041-210X.13906.

The scenario presented in this tutorial is fictional and only intended to contextualize the technical workflow.

About Esri Press

Esri Press is an American book publisher and part of Esri, the global leader in geographic information system (GIS) software, location intelligence, and mapping. Since 1969, Esri has supported customers with geographic science and geospatial analytics, what we call The Science of Where®. We take a geographic approach to problem-solving, brought to life by modern GIS technology, and are committed to using science and technology to build a sustainable world.

At Esri Press, our mission is to inform, inspire, and teach professionals, students, educators, and the public about GIS by developing print and digital publications. Our goal is to increase the adoption of ArcGIS and to support the vision and brand of Esri. We strive to be the leader in publishing great GIS books, and we are dedicated to improving the work and lives of our global community of users, authors, and colleagues.

Related titles

GeoAI: Artificial Intelligence in GIS
Ismael Chivite, Nicholas Giner, and Matt Artz
9781589488441

Advanced Guide to Python in ArcGIS
Dave Crawford and Daniel Yaw
9781589488236

Getting to Know ArcGIS Pro 3.6
Michael Law and Amy Collins
9781589488984

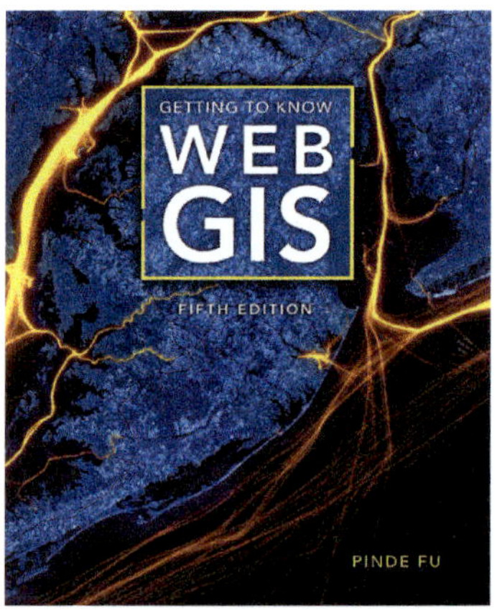

Getting to Know Web GIS, fifth edition
Pinde Fu
9781589487277

For more information about Esri Press books and resources, or to sign up for our newsletter, visit

esripress.com.